Company Commander

IN LOW INTENSITY CONFLICT

Company Commander

IN LOW INTENSITY CONFLICT

Principles, Preparation and Conduct

Major Vivek Chadha

Lancer • New Delhi • Frankfort, IL

www.lancerpublishers.com

LANCER

Published in the United States by

The Lancer International Inc
19558 S. Harlem Ave., Suit 1,
Frankfort, IL. 60423.

First published in India by

by Lancer Publishers & Distributors
2/42 (B), Sarvapriya Vihar,
New Delhi-110016

Printed at Sona Printers, New Delhi.

Printed and bound in India.

ISBN-13: 978-1-935501-14-5 • ISBN-10: 1-935501-14-3

Online Military Bookshop
www.lancerpublishers.com

IDR Net Edition
www.indiandefencereview.com

Contents

Preface	7
Part I : Introduction	9
1. Defining LIC	10
2. Facets of LIC	15
3. Indian LIC Environment	19
Part II : Pre-contact Stage	23
4. Leadership	24
5. Preparations for Move	37
6. Arrival	57
7. Establishing a Post	65
8. First Impact	75
9. Interaction with Paramilitary Forces	82
10. Interaction with Civil Administration	85
11. Psychological Operations	92
12. Human Rights	100
Part III : Principles of LIC	107
13. Selection and Maintenance of Aim	112
14. Offensive Action	121
15. Surprise	131

16. Mobility and Manoeuvre 140
17. Flexibility 150
18. OP Jackal 156
19. Initiative 158
20. Security 166
21. Intelligence 173
22. Communications 185
23. Axis Of Maintenance 190
24. Ground 197
25. Moral Ascendancy 200
26. Media Management 204
Part IV : Tomorrow 207
Annexures 213
Index 233

Preface

This book has been written as a practical guide for Company Commanders in Low Intensity Conflicts (LIC). The book is by no means projected as the way of doing things. It is only a suggested guideline based on my experiences in insurgency in Sri Lanka, Assam, Manipur, Nagaland and Jammu & Kashmir. During six years in counter-insurgency (ci) operations, I always felt the lack of guidance, which ultimately motivated me to write this book.

This is not a conceptual work, which claims to produce the solution for insurgency in the country. It has intentionally been restricted to the practical duties of a Company Commander to help him carry out the tasks assigned to him. I have tried to generalize the suggestions, and not restrict them to a particular area of insurgency, so as to make the study more universal in nature.

The genesis of insurgency has not been touched upon as I felt that it is more important to concentrate on the salient aspects of treaties and documents which are responsible for the turn of events in the particular area. These have been included as Annexures in the book.

During the course of writing the book, I have tried to include the views of as many an authority on LIC as was possible, to give a balanced view to the study. I have also quoted incidents and operations of a period covering more than a century, and of a number of LIC environments, to help broaden the perspective of this study.

I have, while writing the principles of LIC, attempted to initiate a debate and a study to codify the principles, thereby giving LICs the importance they deserve in the present environment. I will be grateful to readers for their valuable suggestions for the improvement of this book.

Part I
Introduction

Defining LIC

Defining and coining a term for counter-insurgency, and related operations, seems to be an operation in itself. The maze of similar terminologies, separated by hair-line differences, have over the years been entirely successful in confusing the reader, and student of military history alike.

To project the problem as it appears — daunting and inexplicable — some of the terms frequently used in this context are Civil Disturbance, Insurgency, Guerrilla Warfare, Subversion, Terrorism, Civil Disobedience, Communist Revolutionary Warfare, Insurrection, Counter-Insurgency, Internal Security, Counter-Revolutionary Operations, Limited Wars, Small Wars and perhaps many more.

The sudden urge to define these kind of operations has stemmed from the variety of uprisings that are taking place the world over. With the change in strategy of anti-establishment forces, from full-scale wars to achieve their means of limited warfare, the need to define specific struggles in specific circumstances perhaps resulted in a number of terminologies. This did not achieve the purpose of gaining instant recognition of the problem but it certainly made its study more 'Mysterious'.

Frank Kitson defines insurgency as, "The use of armed force by a section of the people against the government to overthrow those governing the country at the time or to force them to do things, which they do not want to do."[1] *The support of a foreign country cannot be ruled out to this section of people.*

Hansan W. Baldwin has correctly said with regards to the numerous phrases being used loosely to define this aspect, *"The muddy verbosity and the pompous profundity are beginning to mask the whole subject of counter insurgency and guerrilla war."*[2]

In another attempt at defining LIC, CSM (Company Sergeant Major) James J. Gallagher of the US Army says, *"LIC is defined as a politico-military confrontation between contending states or groups. It is below conventional war and above the routine, peaceful competition. LIC often involves protracted struggles of competing principles and ideologies. Low Intensity Conflicts range from subversion to the use of armed force."*[3] This attempt at defining LIC stresses on the extended duration level and the likely means adopted to conduct it. However it does not suggest the possibilities of moral, military and economic support of a second country, which characterises the majority of low intensity conflicts being fought.

A more comprehensive definition is given by the US Department of Defense which defines LIC as, *"Limited politico-military struggle to achieve political, social, economical or psychological objectives. It is often protracted and ranges from diplomatic, economic and psychological pressure through terrorism and insurgency. Low Intensity Conflict is generally confined to a geographical area and is often characterised by constraints on weaponry, tactics and the level of violence."*[4] This definition has been evolved, after the US decision to classify warfare into three broad heads of "*high*", "*medium*" and "*low*" intensity conflicts. High Intensity Conflicts can be characterised by a World War, where, a number of countries are involved and use of the entire range of weapons can be expected. "Medium" wars are characterised by conventional wars between two countries, examples being the Yom Kippur War and the more recent Iran-Iraq War. Low intensity wars or conflicts owe their classification to the limited use of weaponry and are by no means an all-out war in the true sense of the word. The Vietnam, Afghanistan and Kashmir insurgencies are glaring examples.

However Bernard B. Fall gives the credit of the *"Subject becoming serious academic study and taken out of the hands of the improvisers and sloganeers"*, to the French. According to the author, in 1959 Col Gabriel Bonnet wrote the first history of "Revolutionary Warfare Throughout the Ages", in which he gives it its basic definition. Revolutionary Warfare, says Col Bonnet, *"Consists in the application of irregular warfare methods to the propagation of an ideology or political system."* Dealing both with purpose and method Cols Trinquire and Lacheroy asserted that revolutionary warfare aimed at "...the overthrowing of the government established in a given country and its replacement by another regime ... thanks to the active participation of the population conquered physically and morally by simultaneously destructive and constructive processes, according to precisely developed techniques."[5]

Having seen a number of attempts at defining and describing the most common type of warfare, it is obvious that these protracted and low-key wars are being fought under numerous social, political, economic, military, cultural, ethnic and regional diversities because of which any specific term will limit the scope of any attempt, to study this type of warfare in its completeness. Keeping in mind this factor a single term should be used to make comprehension easier. However, wherever specific examples and issues have to be addressed, terms relating to those subjects are ideally suited. For the purpose of the study undertaken, "Low Intensity Conflict" is appropriately suitable, and adequate, to clear the haze of the war of words and phrases.

Based on this classification and intensity of struggle the 'Spectrum of Combat,'[6] gives a clear diagrammatic representation of the classification of wars (Figure 1).

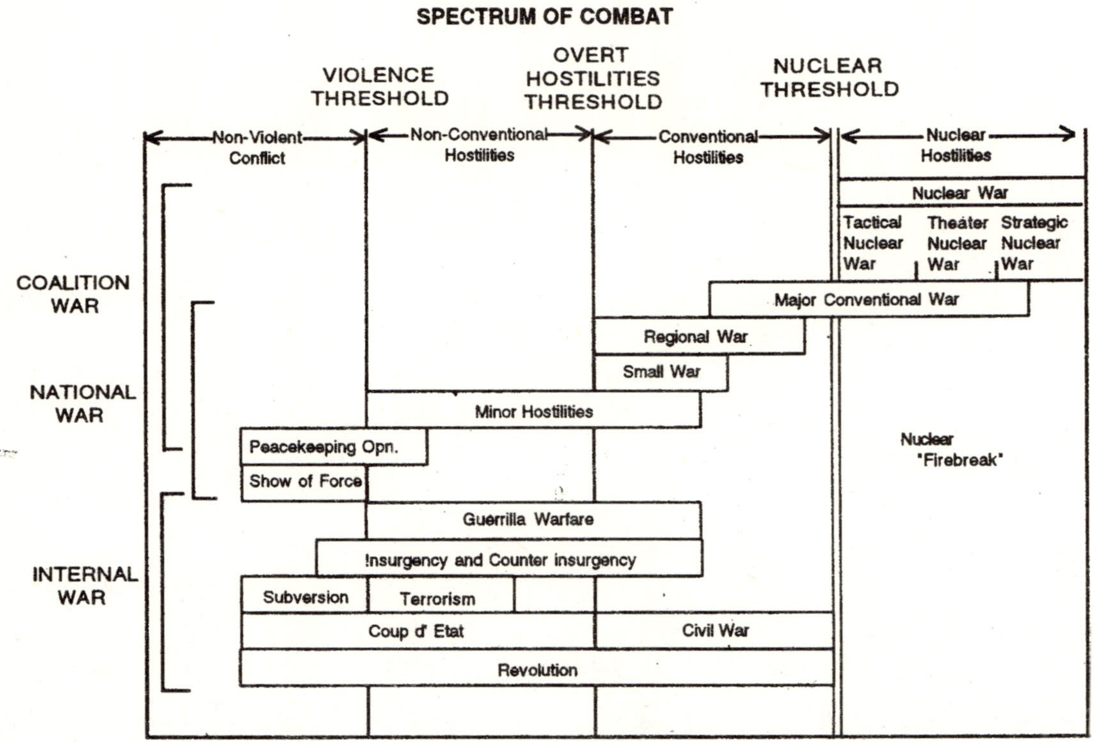
SPECTRUM OF COMBAT
VIOLENCE THRESHOLD
OVERT HOSTILITIES THRESHOLD
NUCLEAR THRESHOLD
Non-Violent Conflict
Non-Conventional Hostilities
Conventional Hostilities
Nuclear Hostilities
Nuclear War
Tactical Nuclear War
Theater Nuclear War
Strategic Nuclear War
Major Conventional War
Regional War
Small War
Minor Hostilities
Peacekeeping Opn.
Show of Force
Guerrilla Warfare
Insurgency and Counter insurgency
Subversion
Terrorism
Coup d' Etat
Civil War
Revolution
Nuclear "Firebreak"
COALITION WAR
NATIONAL WAR
INTERNAL WAR

Figure 1

Notes

1. Kitson, Frank, *Low Intensity Operations,* p. 3.
2. Ibid.
3. Gallagher, CSM James T., *Low Intensity Operations,* p. 3.
4. Ibid.
5. Fall, Bernard B., *Street Without Joy,* pp. 370–71.
6. Dupuy, T.N., *Understanding War,* p. 73.

Facets of LIC

Before the facets of LIC are discussed, let us consider this vital revelation made by Martin Van Clevald. *"Since 1945 there have been perhaps 160 armed conflicts around the world, more if we include struggles like that of the French against Corsican separatists and the Spanish against the Basques. Of those, perhaps three-quarters have been of the so-called 'Low Intensity Variety'."*[1] What is it that has caught the fancy of anti-establishment forces the world over? How has this sudden change in the method of waging war come about? Why have the military campaigns of Alexander, Vegitius, Hannibal, Caesar and more recently Patton and Rommel failed to impress strategists? Perhaps the facets of LIC will answer these questions. It will, therefore, also be pertinent to discuss these, in contrast and comparison with conventional military operations.

It is a well-established dictum that war is an extension of politics. LIC also follows this principle of achieving political ends, by means of a low intensity conflict very much like conventional wars. In fact it is felt by some military thinkers, and perhaps rightly so, that the effective culmination of policies into concrete decisions of international recognition have been more because of LIC than conventional wars. *"Out of the several dozen conventional conflicts waged since 1945, almost the only one which resulted in the establishment of new frontiers was the 1948 one between Israel and its neighbours, and even then the outcome was not an international border but merely an armistice line. During the same period the consequences of LJCs, numerically three times as strong have been momentous. From South Africa to Laos, all over the third world LIC has been perhaps the dominant instrument for bringing about political change. Without a single conventional war being waged, colonial empires that between them used to control approximately one half of the globe, were sent down to defeat through LICs known as 'wars of national liberation'. In the process some of the strongest military powers on earth have suffered humiliation, helping put an end to the entire notion of the white man's inherent superiority."*[2]

LIC, unlike conventional wars, is fought between people of the same country or, at times, by the foreign rulers and the people of the country they are holding. In this case, LIC takes the form of an expression of the feelings of the masses, in their bid to overthrow the perceived unjust oppression. The bid to overthrow the government's forces can also have the active or passive support of a foreign, sympathetic, or interested power. In the Indian context this is the facet all our insurgencies are characterised by.

Another factor which governs LIC is that, force is used to reinforce other forms of persuasion whereas in the more orthodox form of war, persuasion in various forms is used to back up force. The difference hinges on the aspect of 'coup de main'. The difference in classification between the two types of warfare are, the means and the ends.

LIC is never backed by an all-out military effort. This stems from the fact that force applied by the two warring groups is against people of the same country, and any attempt at raising the level of the force can be disastrous for the innocent civilians. In some cases where this factor has been lost sight of, the casualties have been colossal. *"Fifteen years of civil war in Lebanon, a country with a population of 2.5 million, are said to have claimed over 100, 000 dead."* Besides this, *"upto 3 million people are said to have perished during the Nigerian Civil War from 1966 to 1969. Well over 1 million died in the thirty year Vietnamese conflict and another million died in the rest of Indo-China including Cambodia and Laos."*[3]

In LIC the role of the army assumes significance vis-a-vis the civil administration. Whereas in a conventional war, the army concentrates on the military aspect of war, in LIC the army, besides fighting the insurgent, helps and advises the government in the governance of the strife-torn state. The local civil government is marginalised and they perform the duties of supporting the security forces in their tasks.

LIC has been seen as a comparatively cheaper means of waging war to achieve the desired results. This being a protracted means of guerrilla fighting, time and opportunity are virtues of an insurgent. The state of funds can suitably be regulated, by stepping up equipment and heavy arms intensive operations or reducing their intensity. A constant income over a long period suffices in the conduct of this kind of warfare. Besides inputs from sympathetic countries further augment the coffers of the revolutionaries.

The sympathy of the spectators is always with the underdog. In DC the government is blamed for excesses and human rights violations, whereas the insurgents get away even with the most gruesome of crimes. This added pressure on the government helps the insurgents, and thereby increases their popularity. The resultant alienation of the people from the government helps the insurgents.

LIC is perhaps the only kind of warfare, where weak opponents in terms of manpower, equipment, and war support systems, have defeated vastly powerful and superior armies. Going back into the annals of history the British lost India, Palestine, Kenya, Cyprus and Aden, to mention but the most important places where they tried to hold on. The French spent six years fighting in Indo-China and seven trying to keep off embarrassment in Algeria, but after failing in both cases, they gave up the rest of the empire without a struggle, a few minor possessions only excepted. The Belgians were forced to surrender the Congo. The Dutch lost Indonesia, though not before an attempt made to hold on by military means proved hopeless. The Spaniards had the good sense to yield the Sahara almost without a struggle, but the Portuguese in Angola and Mozambique fought for years before they too were forced to capitulate. Even the South Africans, who held out longer than anybody else, ended up by agreeing to withdraw from Namibia.

More recently the Soviet war machine was humbled by the Afghan Mujahideen. *"Inside Afghanistan, opposition to the Red Army consisted of a motley collection of guerrilla organisations. Their members had little formal training, were unable to cooperate among themselves, and never learnt to operate in forces larger than a battalion. Yet nine years later and (the Soviets say) with over 30, 000 men killed in action, the Army was staggering back across the border, jeered by Mujahideen who did not even take the trouble of shooting at it."*[4]

Thus over the years and in the recent past LICs are certainly the most effective means of combat, which if planned and fought, not only achieves the objective but also has the side effect of bleeding the government at times to disintegration as the Afghan Mujahideen did to the mighty Soviet Union.

Notes

1. Cleveld, Martin Van, *On Future War,* p. 21.
2. Ibid.
3. Ibid, pp. 22–23.
4. Ibid, p. 24.

Indian LIC Environment

India provides contrasting avenues of interest, and unfortunately insurgency has also been added to this list. From being limited to the North-Eastern states of the country, insurgency has bounced the further most corners to make its presence felt.

In the country alone and with the exception of India's role in the peacekeeping operations in Sri Lanka, the varieties of LICs have made the task of counter-insurgency operations, an awesome responsibility for the security forces. From at best, the routine tasks being performed by junior leaders, LIC in the country has catapulted them to the forefront of frontline soldiering. The sheer extent of the conflict has made the LIC environment in the country almost at par with a conventional war.

The collusion of some of the insurgent groups specially in the north-eastern states has caused added worry. The understanding between these groups has made their operations better coordinated and the erstwhile advantages of inter-group clashes has slowly faded away. There have been constant reports of the expertise of LTTE (Liberation Tigers of Tamil Eelam) guerrillas in the field of improvised explosive devices (IEDs) noticed in the north-eastern insurgency-hit areas.

Pakistan Factor

Insurgency in India is characterised by support of Pakistan. The defeat of the 1965 war and its partition in 1971 has led Pakistan to seek revenge. The role of Pakistan in materially, morally and diplomatically supporting the insurgent movements is beyond doubt. Robert Wirsing, an authority on Indo-Pak relations, says, *"It is this author's view, set forth in the report prepared for the US Department of State in late 1991, that Pakistan's official involvement in this second round of infiltration was, far from insignificant; that Pakistan supplied sufficient political, diplomatic and material support to the Kashmir uprising; that the material support took various forms, including the training, indoctrination, arming and cross-border movement of infiltrating forces; that the exfiltration of Kashmiri Muslims across the LOC (Line of Control) into Pakistan or Pakistan controlled Kashmir and their covert reinfiltration following training in light arms and guerrilla tactics, played a very important role in maintaining the tempo of insurgency; that the support was planned and coordinated in large part by Pakistan's ISI (Inter-Services Intelligence) and that all this was carried out with the full knowledge and under the auspices of the Pakistan Army."*[1]

Pakistan has also tried to internationalise the issue at every possible forum. The amalgamation of lies, half-truths and vicious propaganda is aimed at defaming the role of the government and the security forces. The virulent rhetoric of personalities no less than the Prime Minister of Pakistan, Benazir Bhutto, says, *"Bullets are ringing in the valley and the people are writing new history. But a new freedom fighter is born with every drop of blood spilling there."*[2]

Role of the Company Commander

In this complicated conceptual and continuing jumble of words, where does the Company Commander find his place? What is his role?

As a soldier who had the opportunity to serve in OP PAWAN, OP BAJRANG, OP RHINO, in insurgency-prone areas of Manipur, Nagaland, North Cachar Hills District and more recently in Jammu & Kashmir, I have no doubt in my mind that LIC or more specifically counter-insurgency operations in our context are operations fought and won by Junior Leaders. These hinge around the Company Commander and take shape through his effective planning, courage and resilience.

The country 'boasts' of contrasting scenarios of rural and urban insurgency environments. It is in the best interests of a Company Commander to study them and adopt his line of action likewise. Most of the North Eastern States present rural insurgency areas with the exception of medium sized towns and cities. On the contrary, Jammu and Kashmir offers a mix of rural and urban areas and to make matters worse it also leads to operations in contrasting terrains.

Most of the readers, I am sure, are well aware of the genesis of insurgency in the Indian context. Keeping in mind the present state of these uprisings, I will like to emphasize on certain vital documents and treaties which have moulded the history of insurgency in the country. These documents are the turning points in the lives of the people and may prove to be the future-guidelines for any settlement.

I will like to emphasize on Jammu & Kashmir and the NSCN-affected areas of Nagaland and Manipur while projecting these treaties/documents to the reader. In the case of Jammu & Kashmir the following treaties/documents require careful study by Company Commanders to understand the problem in its correct perspective:

- Instrument of Accession and reply by Lord Mountbatten (Annexure I)
- Resolution of the United Nations Security Council of August 13, 1948 (Annexure II)
- Shimla Agreement of July 3, 1972 (Annexure III)
- Kashmir Accord (February 1975) (Annexure IV)
- Article 370 of the Constitution of India (Annexure V)
- Armed Forces (Special Power) Act (1958) (Annexure VI)

As regards insurgency in Manipur and Nagaland, the 10 year Akbar Hydari Agreement June 1947 (Annexure VII) Shillong Accord signed on November 11, (See Annexure VIII), are the more relevant documents which merit study.

Notes

1. Wirsing, Robert G., *India, Pakistan and the Kashmir Dispute,* p. 119.
2. Jagmohan, *My Frozen Turbulences in Kashmir,* p. 403.

Part II
Pre-contact Stage

Leadership

John Keagan, describing the elements of leadership says:

They have comprised an element of "kinship," by which the leader surrounded himself with intimates identifiable by his followers as common spirits with themselves... which has been bolstered by 'sanction', the reward or punishment of followers... which has been reinforced by "example," the demonstration of personal acceptance of risk by authority... example has been amplified by prescription' of the explanation of the need for risk-taking by the leader... which has culminated in 'action', the translation of leadership into effect.[1]

During the course of approximately six years in the LIC scenario, I feel that in this environment leadership has a very difficult and challenging task of continuously motivating and pushing the troops against an unpredictable, invisible and irresolute insurgent.

Leadership is a process of persuasion by example by which the leader induces his command to pursue goals set by the leader or shared by the leader and the led.

So much has been said and written about leadership that it has become a jumble of words, adjectives and idealistic principles. Most of it is generic and does not correlate leadership with specific tasks. The leadership qualities for a statesman, politician and soldier will differ. Therefore among the leadership traits certain qualities will gain prominence over the others for a specific scenario of LIC and specially so for a Company

Commander as he assumes the role of a fighting commander and an able administrator.

Keeping in mind the peculiarities of LIC I shall now examine some of the desirable qualities of a Company Commander.

(a) Initiative

We have been brought up on peacetime routines which do not demand initiative as one of its primary requirements, and specially so in the case of junior leaders. Caution at times is preferable to action. However contrary to this working philosophy, *LIC requires initiative to seize what does not exist, see what seems invisible, scent, perceive from the faintest of indications, and chase the impossible because the restless and indomitable spirit of initiative fails to fail.*

The task of a Company Commander does not end with his initiative. His success is in his subordinates imbibing the quality.

There will be numerous situations, when after a number of failed attempts, opportunity will present itself to the least expectant company commander. Here the initiative to seize the threads of information to build the operation will make the difference between success and failure. This is what Lt Gen SC Sardeshpande has to say about initiative based on his experiences in OP PAWAN. *"In war the only rescuing factor is to do, to act, to take the first step (initiative and action may prove wrong on occasions, but their chances of success and maintaining cohesion too are far brighter than inaction, awaiting orders or delay in doing a right thing).*

Young officers were fairly prompt in doing, acting and doing things right. They did not wait, for doing right things; and therefore largely succeeded. What the young subaltern lacked in knowledge and experience he more than made up by enthusiasm, keenness, guts and getting into the act promptly. God bless him."[2]

(b) Lead by Example

In LIC there is no scope for detaching oneself from operations. The conflict is a continuous day and night cycle of events which will require guidance and policy decisions to be given. Trying to give a loose rope and calling it encouraging only proves disinterest and callousness on the part of the commander. There is no wisdom in being so far from the scene of action that one misses the essential ingredients of action.

This poem perhaps sums it up well.

"Good Morning, Good Morning! the General said
When we met him last week on the way to the lines,
Now the soldiers he smited at are most of 'em dead,
And we are cursing his staff for incompetent swine
He's a cheery old card, grunted Harry to Jack,
As they slogged up to Arras with rifle and pack.
But he did for them both with his plan of attack."

In operations which go on days at a stretch, unless the Company Commander is there, with his command, rallying them and leading the occasional patrol, the troops are bound to lose respect for the leader. In the vagaries of battle the presence of the Company Commander will produce the necessary electrifying effect to snatch *"Victory from the jaws of death"*.

The Indian Army has a rich tradition of the leader leading by example. This has been amply illustrated by recent conflicts in LIC areas during OP PAWAN. The share of officer casualties was 20% (both dead and wounded). We have sustained this tradition at a very heavy cost. Now the time has come that acts of valour have become traditions. It is expected that a Company Commander will lead. And this will take the Company Commander on a pedestal which is reserved for the hero of the troops, an elevated position which can only be achieved by being there right up in front.

(c) Courage

A Company Commander and a leader needs courage in good measure to goon relentlessly against a resolute insurgent. It is not just bravery for a particular incident, but sustained willingness to take risks to function effectively under prolonged stress, to survive defeat and keep going on that defines courage.

Perhaps LIC is the most realistic test of a Company Commander's courage. It is this pivot which is responsible for operations against all odds. Despite failures, death, injury and prolonged stress, it is the indomitable courageous stand of the Company Commander which motivates the men to hit back and regain their confidence and morale. Let us take this incident of a newly-posted company commander in insurgency who had lost a soldier in an ambush.

The solemn looks on the faces of the soldiers of 'A' Company betrayed their bleeding hearts and tear-filled eyes. The company sat in front of their Company Commander looking at their newly-posted leader. The disheartened and defeatist look on their faces was the direct impact of the death of their young comrade in an insurgent ambush. The bullet-riddled body, a peaceful and stoic symbol of death had shaken the Company. The Company Commander faced the awesome task of rejuvenating this grieving band of soldiers, to whom death had come as a rude reality of life. He knew that his spirit and courage of conviction will make or mar the future performance of his Company. He got up and looked straight into their eyes.

The faces of the soldiers lifted and their unsteady gaze rested on their leader, "It is unfortunate that I stand here today in my first meeting with you all to mourn the death of our brother in arms. I know the grief of this separation is unbearable, that a young soldier has been snatched by the cruel hands of destiny and that his loss is irreparable." The Company Commander paused for the words to sink in and then in a louder and clearer voice said, "But remember that death is a festival of war, we have all come here for a purpose, we have all been trusted with the responsibility to bring peace in this war-ravaged land. Sanjay is fortunate to have laid down his life in the true traditions of a soldier. Do not shed tears, because you do not pay homage by these to the departed soul; do not lament because you only make death lose its meaning for Sanjay. If you want to show the brave soul that you love him, then wipe off these tears and swear that you will avenge his death by fighting insurgency with a fire that will burn in our hearts till we stay here. I hereby swear that in the next fifteen days we will hit back and bring glory to the Company and Battalion. It will be our way of saying that 'Sanjay, we salute you'. The meeting ended with a two minute silence. The very next week, the company got its first success in the form of an HM (Hizbul Mujahideen) insurgent who was killed during an encounter.

(d) Patience

Insurgency is a battle of nerves and whoever gives in first ends up the vanquished. There may be a number of situations where a patient and steadfast approach ultimately yields results.

The insurgent will always be as slippery as an eel. Insurgency as discussed earlier is a fight against an invisible enemy. The Company Commander should be patient in his approach and should not jump the gun. Opportunity will not be kind enough to present itself again. The blow should be delivered when the chances of success are substantial.

Sources prove to be another 'species' which test one's patience. The Company Commander must understand that he is dealing with an insecure and vulnerable man who has the constant fear of being exposed. The lack of patience and understanding will not only jeopardize the entire operation, but it can also endanger the life of the source and your own men.

After reaching the area of responsibility one generally realises that the population is anti-establishment and anti-army. Repeated attempts at trying to tell the futility of insurgency often fall on deaf ears. This is a common cause for losing patience and at times results in excesses. The unprovoked actions of a Company Commander give a loose rope to the company and result in situations which prove to be embarrassing to the army and the government.

Another link in the chain is the soldier. He stays under constant tension and pressure. The difficult and often dangerous life style of LICs causes an occasional vent to pent-up emotions. An understanding and patient Company Commander is required to solve the problems and control the situations.

LIC is characterized by colossal pressure from higher headquarters. These are momentary flashes and do not actually demand rash actions or decisions. But during this period of stress, patience helps to stand up to the situations and yet produces results in suitable time frames.

(e) Morality

Some time back, going through the magazine section of a daily I came across an interesting article which had covered a reception for the brave soldiers of the Second World War by the Chief of the Army Staff. Colonel Malik, a war veteran of the erstwhile Baluch Regiment, relating an incident of a Prison of War Camp said, *"After the war, I was with a PW Camp, in which there were three or four officers of the rank of Maj Gens and Brigs. One day the Gen had tears in his eyes. When I asked Lt Nakamure why the Gen was upset, his answer became the most memorable thing that lives on in my mind. He said that the Gen was upset because he had learnt that his men had been stealing rice on account of insufficient rations given to them. He was worried what effect such actions would have upon the character of his men, and not about the insufficient rations they were being supplied."*[5]

In a scenario where the Company is constantly being exposed to the civilian style of functioning, a style hithertofore unknown to the soldier, unless a high standard of morality is maintained, principles, truth and ethics are often sacrificed for petty gains. For the Company Commander it is imperative to prove his morality to the company, and for the Company to the people.

LIC is accompanied by many situations which force lies and double-faced actions to be taken for the good of the organization and cause. It has rightly been said that "war is a dirty game". But the Company Commander must draw a line between professional attempts of knowingly crucifying truth and morality on the alter of organizational betterment, and contrary to it for personal benefit. It is not only important to know it as the Company Commander, but perhaps more important that the Company understands it and appreciates it.

In insurgency where the insurgents take the support of deceit, morality serves as a positive indicator of good intentions and undiluted concern. This can be proved to the people by obviously visible actions to get an upper hand over the insurgents.

(f) Believe in the Fight

It is not only difficult but impossible to put up a false front of believing in what one fights for, when feelings do not substantiate the belief. The Company Commander must believe in the cause and only then can he expect his Company to put in their very best. For the insurgent it is very easy to, play up the sentiments of the people and push them to the brink of death to extract his motives. This is one characteristic which is best learnt from successful insurgencies like Vietnam where the insurgent leaders were able to unite the people and fight as a whole. *"Our war of liberation was a people's war, a just war. It was this essential characteristic that was to determine its laws and to decide its final out come."*[6]

Similarly the insurgents in El Salvador have proved formidable opponents to the government forces because of their belief in their cause, their belief in their fight. Despite numerous tragedies and sufferings they continued relentlessly. A poem written by a revolutionary sums it all up.

(g) All Kinds of Reasons

Here you find me:
Just completed my thirty-second year.
a widower twice over.
I have fungus
and dandruff
and a rotting molar, in my left ear
I am deaf,
But no bullet has touched me yet.
I've swallowed dust,
mud, fog,
I've chewed stone
with my teeth,
but I've fought at the side
of the struggling multitudes
at their barricades,
Embraced by the fire of the revolution
in this canyon,
uphill,
down hill,
loving as I love this life, poetry,
and the people,
how can I not offer up my bones?[7]

The company has to be made to understand that the fight

against insurgency is a fight between the government and the anti-government establishment. The rich history and gallant traditions have for years proved that victory comes to the soldier who fights for a just cause. In this fight too which the Regiment, Battalion and Company finds just and noble, his contribution will achieve the cherished aims of the nation.

But remember, you as a Company Commander must first believe in it.

(h) Humour

LIC presents the soldier with an extremely difficult life style of protracted and long-drawn-out battle. A Company Commander can make it worse by building up pressures for an assigned task. Contrary to this the entire work culture can be altered by a light and healthy atmosphere by a humorous Company Commander.

Humour as a characteristic of the Company Commander brings in informality which is required to create an atmosphere of constructive leniency. This serves as a catalyst and bolsters the performance of the Company. It has been rightly acknowledged that "humour adds spice to life", and spice adds flavour and the zip to make things tick.

Humour is also an effective tool for reducing tension as this incident illustrates. *"An American platoon was pinned down by heavy enemy fire. The Lt in command was trying to take a head count of his men who were paralysed by fear. 'Williams where are you?' he called. 'I'm here', replied Private Williams who was cowering behind a boulder which was being sprayed by enemy bullets. 'Stand up, so that I can see you, said the Lt 'Lt', called back Williams, 'If you want to see me, you stand up', causing laughter and dissipating fear."*[8]

(i) Accept Responsibility

This quality of a leader has two connotations and both are equally applicable to a Company Commander in LIC. The first is the willingness to accept responsibility for a task which everyone is keen to avoid. This is linked with the quality of decision initiative, but it also adds burden and bearing the conseqences. But David Keppel's famous poem perhaps contradicuts what I say.

Better never trouble trouble
Until trouble troubles you
For you only make your trouble
Double trouble when you do
And you trouble like bubble
That you're troubling about
May be nothing but a cipher
with the rim rubbed out.[9]

The second aspect is being accountable for your reactions. Unlike war where there is no time for either the Company Commander or for the Company to weigh, contemplate and discuss decisions taken during combat, LIC offers adequate time after the action to do so. The Company Commander is and should be accountable for his decisions and actions. Therefore, before you take a decision it is better to weigh it rather than regretting its consequences.

(j) Professional Competence

Professionally incompetent decisions in any sphere of service life are unpardonable because these stem from callous indifference, casualness and contempt for the system. A professionally incompetent Company Commander will not only get killed in LIC but worse he will get the men he commands killed.

The lack of leadership traits makes a Company Commander

incapable of taking sound tactical decisions. This will adversely affect the performance of the Company and in turn its morale and effectiveness. This will also lead to the Company Commander failing in his administrative and technical duties which will surely reflect on the company's capability to fight.

An incompetent Company Commander cannot have the moral courage to correct and teach his command. The fear of so exposing his inadequacies will stop him a step short of doing so.

The Company Commander cannot think of leading by setting an example because of the lack of professional competence. The confidence of *"Let me show you how it is done"* will be non-existent.

(k) Confidence

Not in the clamour of crowded street,

Not in the shouts and plaudits of the throng,

But in ourselves are triumph and defeat.

— Henry Wordsworth Longfellow

LIC is an unpredictable fight which hinges itself on the confidence of the leader. It is only a confident and self-assured Company Commander who can take decisions when things do not look bright, when defeat is imminent and when all but a few have given up. It is during these circumstances that a Company Commander must weigh and study the alternatives and take a decision confidently.

Confidence like fear is infectious. The key to a confident, motivated and self-assured Company, is a confident Company Commander. It is therefore important that a Company Commander remains self-assured, confident and imbibes this feeling in his troops.

During the course of everyday combat, the Company

Commander will be required to take decisions, risks and initiative. All these qualities can only surface when the Company Commander has confidence in himself and in his command.

(l) Physical and Mental Resilience

These two qualities when combined form an irresistible combination which is capable of achieving any standards set for a leader. An unknown poet has beautifully described these qualities.

When things go wrong as they sometimes will,
When the road you're trudging seems all uphill,
When the funds are low and the debts are high,
And you want to smile, but you have to sigh.
When care is pressing you down a bit,
Rest if you must, but don't you quit.
Life is queer with its twists and turns,
As everyone of us sometimes learns.
And many a failure turns about
When he might have won had he stuck it out.
Don't give up though the pace seems slow,
You may succeed with another blow;
Success is failure turned inside out –
The silver tint of the clouds of doubt,
And you never can tell just how close you are,
It may be near when it seems so far.
So stick to the fight when you're hardest hit
It's when things seem worst that you must not quit[10]

There will be situations when the mind will just about give up in LICs. The protracted nature of warfare preferred by the insurgents requires mental and physical resilience to go on trying, because in this battle of nerves' it's either the insurgent or the soldier who gives up, and that makes the difference between the victor and the vanquished.

(m) Vision

Above all, the Company Commander must have the capacity to deduce from the available inputs, the activities for the future. Rather than being a man for today, he should be able to think for tomorrow and the Commander who lacks vision cannot plan for tomorrow. In LIC battles may be won or battles may be lost, but ultimately it is the war against insurgency that has to be won — besides there are no runners-up in war.

In short, with respect to insurgency, the future announces itself from afar. But most commanders are not listening. The noisy clatter of the present drowns out the tentative sounds of things to come. The sound of the new does not fit old perpetual patterns and goes unnoticed by most commanders.

Notes

1 John Keegan, *The Mask of Command.*
2. Lt Gen SC Sardeshpande, *Assignment Jaffna,* pp. 149–50.
3. Sugfried Sassom, *Collected Poems.*
4. Lt Gen SC Sardeshpande, *Assignment Jaffna,* p. 141.
5. Subhra Mazumdar, *The Hindustan Times,* Aug 30, 95.
6. Vo Nguyen Giap, *People's War People's Army,* p. 49.
7. Jon Lee Anderson, *Guerrillas,* pp. 58–9.
8. Brig Nazareth, *Psychology of Military Humour.*
9. Norman Vincent Peale, *Treasury of Courage and Confidence.*
10. Ibid.

Preparations for Move

It is common knowledge for a commander at any level of combat, that a well-prepared soldier is a well-motivated soldier. Besides the factor of motivation, preparation or the lack of preparation directly affects the morale, confidence, will to win and the elementary mental and physical fibre of a soldier.

It is desirable that a multi-faceted schedule for training is planned, and executed keeping in mind the operations. It is imperative that a soldier is not pushed into battle blindly, and made a scapegoat because of the lack of forethought, planning and spadework in the preparations for move. The experiences of the army in a number of fire brigade actions have an unenviable record of misadventures. The words of Lt Gen KS Brar in his book *Operation Blue Star, The True Story*, sum up the level of preparedness of the army before the operation:

"Very scant information was forthcoming from local police and intelligence agency sources... In fact, most of it was woefully inadequate, unconfirmed and outdated. The estimates of the number of extremists entrenched in the Temple complex were at wide variance and more the result of speculation rather than of hard, accurate intelligence."[1]

Though a 'general' and vague briefing does give the company commander the credit of giving a loose rope to his subordinates, but it has more often than not resulted in their getting hung by it.

The lack of preparation for any operation can have disastrous consequences. Examples of these fiascos are not difficult to find even in our day-to-day pressure-free peace-time routines. An incident which took place in my battalion not too long back clearly shows what a general officer-like briefing can do.

An exercise had just got over, and the battalion had acquitted itself with the customary clockwork efficiency of a well-oiled fighting machine. The Commanding Officer (CO) had called all the ladies to the exercise location for a get-together. The responsibility for making the arrangements for the get-together were given to a budding officer. Since the officer wanted to personally supervise the arrangements, he called a JCO from his company and giving him a list of items to be purchased, sent him to the market. However, an addition to the eatables was given by the Adjutant at the last minute to the officer saying, "Sir, why don't you add patties, The CO's wife loves them." The officer thanked the Adjutant for his valuable suggestion and told the JCO, "Sahib [a term of address connoting a degree of respect] list wale samaan ke saath-sath, patties le ana." (Saab, besides the items mentioned in the list, do get patties also.) However, the officer, not used to ensuring the correctness of understanding, and neglecting the level of awareness of the JCO, did not add it to the list with the JCO.

The JCO bought all the stores as given in the list except the one added orally. He perhaps mispronounced the item and was sent to the neighbouring shop selling clothing stores. His requirement was immediately fulfilled here and he returned to the battalion satisfied.

The officer took all the eatables from the JCO and dispatched him after thanking him. The eatables were being checked by the officer when the CO came to see that the arrangements were up to the mark. Having been satisfied with the 'field' preparations, he suddenly asked the officer, "What do we have for snacks?" The officer, wishing that the CO should ask this question replied, "Sir, besides pastries and sandwiches, I have also arranged for patties." "Lovely," replied the CO." "Let me see, I hope these are chicken patties, my wife loves them." The officer sprang and gave the packet to the CO. The CO opened the packet and pushing his hand in it took out the supposedly mouth-watering contents. Out came a pair of panties that the JCO had brought for his 'saab'. His pronunciation and understanding had given his unsuspecting and upcoming company commander a shock he had least expected, and the CO adequate "food for thought"!

(a) Psychological Preparation

Blessed is the match consumed in kindling flame

Blessed is the flame that burns in the secret, fastness of the heart

Blessed is the Heart with strength to stop its beating for honours sake

Blessed is the match consumed in kindling flame.

– Hanna Szenes[2]

Parachuted into Hungary in 1944, officially to organise the rescue of downed R.A.F. (Royal Air Force) pilots, and unofficially to save Jewish refugees, captured and shot by Hungarian Fascists shortly before the end of the war, Hanna lit a fire in the hearts of the Israeli paratroops, who memorise her poem.

What is it that drives a soldier to his death in the face of the enemy? A number of reasons can be attributed to this quality. Respect for the clan, religious zeal, 'Izzat', love for the colours, blind faith in leaders, following the tradition of his forefathers, plain call of duty and the love and comradeship amongst soldiers can be some of the factors acting as the driving force. But ultimately one or a combination of these are galvanised by a good, capable and motivated Company Commander to achieve the desired level of psychological advantage over the insurgent. It is absolutely essential to achieve this in order to drive the men against the motivated and often frenetic insurgents day in and day out. Though this is a continuous process, the foundation for the same has to be laid in the preparatory stage before the move.

The Company Commander must also ensure that all available information about the proposed area of operations is given to the Company right down to the junior-most soldier. At times, it would be prudent to repeat the information being shared to ensure that all personnel are able to listen to it despite the daily routine of duties and state of constant flux. The ideal course will be to only give as much as the Company can digest. Even this should be disseminated during relatively informal sessions rather than at an official classroom scenario, which is more of an exercise in futility rather than a forum for sharing information.

The Company should also be told that unlike a conventional war situation, they are by no means fighting the 'enemy' in the true sense. Though the line dividing an insurgent and enemy is very thin, the basis of fighting insurgency is to recognize and appreciate the significance of this line, because the war against insurgency is more often won by winning the insurgent rather than by killing him.

Very much like a conventional war, the area is strife with rumours and propaganda. Though some of it may be incidental, most of it is the creation of the overground sympathisers of the insurgents who work overtime to convert the public at large for the cause of insurgent movement and to erode the morale of security forces. In this war of attrition, it is often the side which is more motivated, and makes better use of propaganda that wins the hearts and minds of the people: the basic requirement for victory. After all, defeating an insurgent requires conquering his mind. This can be achieved after the surrender of his mind and the surrender of his will against aggression. In the two-pronged strategy, the Company Commander will have to make the idea of surrender of the will for his company unimaginable. Though the methodology can be varied, the results should be as blatantly obvious as the attitudes of the Japanese and the Germans towards war in general and death in particular during the Second World War. On the other hand, the idea of surrender has to be more profitable than the miseries and harsh realities of the life of insurgency for the insurgent. The effect of concerted, well-thought of and systematic propaganda can make the area of responsibility army friendly from being insurgent-friendly. One such example of so called turncoats, in the insurgents' language, was of Chilly, an insurgent stronghold in the Doda District of Jammu & Kashmir. This is what the insurgents of the Hizbul Mujahideen had to say about the people of Chilly. The original and translated versions tell the effects of shrewd propaganda.

।। जरूरी इल्लहि गज़ल ।।

हमें गया किरेगा चड़ढ़ा हमें क्या किरेगा चड़ढ़ा

1. हमारे साथ है खुदा हम सिपाही है खुदा के हम उमती है रसूल के मुस्तफा कबीले हम पर निगाह
2. वह वख्त आने वाला है वह घड़ी आने वाल है (कश्मीरी)
3. नारे तुहीद छोड़कर
   कलमा तुहीद छोड़कर
   चिल्ली के गद्दारों तुम पर लानत है।
4. ऐ चिल्ली के गद्दारा तुम ने चढ़े
   कश्मीरी तुम्स छुई इनाम
   यह चिल्ली इलाका वालों की सेवामें
   हाजिर है।

(b) Induction Training

The more you sweat in peace the less you bleed in war

— N.D.A. * Weapon Training Area

Train Hard, Fight Easy

— S.A.S. ** Training Wing

To lead untrained men is to throw them away

— Confucious Analects XII (c. 500 B.C.)

A soldier knows that death, sacrifice, injuries and hardship are daily companions in war: only good, sincere, realistic training or an end to fighting alleviates them. But soldiers unlike their commander do not carry the burden of responsibility in combat. You as Company Commanders face and will continue to face the joyous smiles of your command after a job well-done and the moist eyes and heart-rending scenes of casualties after an encounter, a blunder. It will be this particular appointment amongst a number of others in the army that will bear the burden of explaining the irreparable loss of life to the command and more importantly to oneself. And their lack of training will stand out as the single most important reason, time and again, reminding us of our failure to look after the welfare and safety of the men we command.

As Company Commanders you will never be able to muster up a substantial strength of your company despite clear orders for mobilization with adequate notice. The all-too-familiar parade state tells the harsh realities of "unavoidable guard duties", "pressing leave problems", "immediate hospitalizations", "inexplicable temporary duties" and last but not the least the inescapable central duties. Despite these uphill and supposedly insurmountable problems, training must commence and training

---

* National Defence Academy

** Special Air Services

must go on because induction will certainly not be delayed. However, never curtail leave because in today's environment, it is not difficult to find reasons to curtail leave, which is nothing but poor planning and the sign of a weak-willed commander. The problem of non-availability of manpower can be solved to some extent by the following means:

- Divide the training programme into essentials and add ons.
- Run the essentials more than once by recycling the unavoidable duties and responsibilities.
- If time permits, re-run the add-ons.
- If some personnel of the Company are still left out, do not hesitate to make them attend the missed class with some other Company.
- Ultimately, keep trying for the availability of your manpower.

The Company will normally be tuned to the methodology, drills and procedures of conventional warfare which are very different from the unconventional, inexplicable, at times insurmountable and quite often unethical fight in LICs. It is the single most important task of the Company Commander to willingly erode the ingrained basis of conventional war and painstakingly imbibe the vastly different tenets of LIC. It will initially be very difficult for the simple guidance-hungry soldier to readjust to the war of attrition where initiative, ingenuity and mental flexibility form the basis of survival. These are the willingly-suppressed qualities which will have to be nurtured and allowed to flourish at times even at the cost of traditional military standards of peacetime soldiering. Havildar B.H. Mujhawar is one such striking example. After days of unfruitful operations, we had ultimately caught the long-absconding brother of one of the PTMs (Pakistan Trained Militants) in our area. We suspected that he had hidden an SLR (Self-Loading Rifle) which was part of the booty taken from the site of a BSF (Border Security Force) ambush by the militants. We tried our level best, but could not find the location of the weapon. Havildar Mujhawar who was standing at the door was also privy to the interrogation. Tired of our routine questioning, I tried to weigh down the young boy by telling him that I would kill him if he did not tell me the truth within 10 minutes, the time given to him to mull over the option.

After 10 minutes Mujhawar brought him back and before I could say anything, smiling from cheek to cheek said, "Saab, the rifle was never with him, but he has consented to show us the location of two boxes of explosives, ammunition and other warlike stores ofthe HUA (Harkat-Ul -Ansar)." I was shocked, but impressed with the results: obviously the NCOs (Non Commissioned Officers) line of action was more effective and realistic than my own.

Some features for conduct of training that are suggested to be borne in mind are:

- Besides the basic theoretical classes, it is important to let the junior leaders experiment and enforce the drills learnt, so that their confidence in the efficacy of the drills is increased.
- It is certainly more important to do only a part of the training syllabus correctly rather than completing the training, yet coming out of it with utter lack of confidence.
- The training should preferably be carried out in the terrain similar to the terrain of the projected area of operations.
- The training should be carried out in four 'shifts', i.e. there should be four time slots of six hours each and these should be shuffled to ensure that even during routine days training, the ability to work at any time of the day or night is attained.
- Protracted exercises should be conducted over three days to increase the mental and physical resilience of the company.
- The company should be trained in fixed sub-units based on

the availability of manpower rather than the paperposted formality of sections. Duties and tasks should he allotted to each sub-unit and platoon.

- The buddy system of functioning should be enforced at the permanent location itself to imbibe cooperation, understanding and affinity amongst the pairs.
- Use of IWESS (Infantry Weapons Effect Simulation System) equipment can be a very good means of finding out the standards achieved by the sub-units and will also give a sense of achievment and fulfillment to the soldiers.
- All the weapons must be tested for their functioning. This not only helps in their effective use, but it also helps in building the confidence of the company personnel in their weapons.
- Special emphasis should be given to improvised drills such as the ability to fire from both the shoulders, hip fire techniques, night fire, fire while crawling, side rolling and jumping.
- Innumerable improvised styles of operations have been developed by individual units during their stay in CI operations. These are effective and keep the insurgents guessing. Some of these will be discussed subsequently when the case studies are taken up. The company can be explained these variations, and practised with these styles.
- The effect and use of all weapons and equipment must be shown to the Company. This goes specially for the much-maligned and often discarded weapons like hand grenades. A case described may change the opinion of most of the readers as regards the effectiveness and utility of this 'Life Saver'.

The freezing cold chill of the ice-covered mountain was enough of a dissuading factor for most patrols in the winters, thought the insurgents were sitting in the Gujjar 'gwari'[3] enjoying their meal and warming their battle-worn bodies. A lonesome sentry stood sentinel to the group of five HM cadres least expecting trouble.

A patrol led by a young energetic JCO (Junior Commissioned

Officer) on receiving information of the group was moving stealthily towards the 'gwari'. The steep climb and waist-deep snow had made progress a slow, tiring, painful and tedious process. At times the weapons had to be used to push the body up from the frozen expanse of snow.

As the patrol neared the 'gwari', the JCO checked his equipment and felt the dead weight of his ammunition and grenades. He remembered the words of his strict Company Commander, "Saab, kabhi bhi ammunition our samaan mein kanjoosi mat karo, ek din yahi aapki jaan bachainge."

On closing to the 'gwari', the JCO surrounded it and he along with two men closed in at the entrance. The crawling bodies of the freezing soldiers made slow but steady progress towards the 'gwari'.

Suddenly the sentry noticed the JCO and raised an alarm. Left with no choice, the JCO fired his carbine. The first shot hit the sentry on the leg but before the JCO could reach him and capitalize on it, his weapon would not fire. The weapons of the other two soldiers too fell silent magnifying the stillness of the eventful night. The effect of snow and low temperature had taken a toll of the weapons. The JCO as if in slow motion saw the hand of the insurgent grab for the fallen rifle. Each jab of pain, and the fear of death was evidently visible on the face of the struggling insurgent. The ominous sign of death loomed large and vivid in front of the three hard-pressed souls. As if by an involuntary action, the JCO struggled with the grenade pouch, and pulling the grenade free from the pin, lobbed it at the insurgent.

The blast of the grenade shattered the stillness of the night and acted as the life-saving thrust before the militants were eventually killed. The prophetic words of the Company Commander rang clear and loud in the JCO's head.

— Aim at variety in training by making the program interesting to comply with. Often the teacher and the taught feel that too many features or too much of a particular subject has been pushed through. Video films are good means to add spice to the training programs.

— Competitive spirit is the basic nature of a human being. This should be exploited by conducting inter-platoon or even inter-combat section competitions after each leg of training to induce involvement and interest in training.

— Be where the action is. Experience has shown that the standards achieved under the watchful guidance of the Company Commander are higher and have also helped in inducing the spirit of vibrance and morale of the Company Commander.

— It is profitable to train interpreters from men of your Company, if time is available. Incidents of officers and JCOs taking to local people and asking about the LTTE but being unable to understand the location despite the people actually giving the location of the LTTE was not very uncommon in Sri Lanka.

(c) Intelligence Collection

It is a well-known fact that the key to LIC is intelligence. This is primarily of two types: background or generic; and specific or actionable. I will deal with the aspect of background or generic information alone, as actionable intelligence is neither valid, nor of any use in the preparatory stage.

For the Company Commander and the company, it is important to have basic levels of information. The details that you may like to know and feed your Company in the form of informal discussions and training programs can be:

— Which the general and if possible specific area of operations is.

— What the nature of terrain, climatic conditions and vegetation is.

— What the demographic pattern of the area of operations is.

- What the setup of the civil government and paramilitary forces is.
- What the genesis of the problem, its causes and level of support from the people is.
- How effective the movement has been.
- Which the major groups, their ideologies and weapons and equipment being used by them are.
- What the modus operandi of the militant groups is.
- What the modus operandi of the security forces has been.

Besides getting this basic information it is also a useful and a time saving exercise to start working upon the task in hand to give a systematic approach to intelligence collection. The following actions are suggested:

- A local newspaper should immediately be subscribed to and relevant cuttings should be filed. This, besides giving relevant information also helps build a broad framework, and widens the perspective of the problem at hand.
- A diary of events should be made which has the details of day-to-day activities and also the information gained. A column should also be give your deductions and action proposed to be taken.
- It would be ideal in case a list of militants of The area can be collected from the formation HQ. This gives the much-needed tangible target to work upon and also helps avoid spadework when time is fast running out. Here I would like to cite the example of OP BAJRANG. This operation was launched with the bare minimum of intelligence and shockingly even the essential knowledge of knowing the names of the insurgents was lacking. The usual exercise for any of the Company Cmmanders was to stop people and ask as to who were the leaders of ULFA (United Liberation Front of Assam) in the area. This process alone took a couple of months to complete and at times, the innocent-looking informer was the much-relieved

insurgent pointing towards his own house. And the so-called 'Top Ten' insurgents on our list during this period later turned out to be village-level cadres to our embarrassment. Our confidence and the confidence of the troops took a knocking in these initial days in the area.

— The size of maps prepared need not only be limited to the area of responsibility of the company. In fluid operations of ever-changing locations, pattern-building is a systematic approach of predicting the location of the militants.

(d) Physical Conditioning

Torture, waste of time, disgusting, pain in the neck, useless, etc. are some of the adjectives and words we use when it comes to pushing ourselves physically to attain a high order of physical fitness. How many of us realize that development of the heart and muscles builds our bodies which in turn improves our physical strength and mental well-being, the basic ingredients for a soldier to perform the tasks assigned to him. Let us hear our counterparts in the Russian and German armies:

"Physical training and athletic competitions permit people to develop endurance, self-mastery, courage, and even competitive aggressiveness."[4]

"Sports are an indispensable element of training and education in the Bundeswehr."[5]

Physical fitness not only gives satisfaction to a soldier, but it also gives confidence to the commander in the ability of his troops to perform the improbable if not the impossible.

Contrary to belief, physical fitness not only concerns itself with the physical well-being of the body, but it also goes a long way in enhancing the mental flexibility and mental alertness of the individual. It is only a physically fit soldier who after undergoing extreme physical exhaustion still has the strength to stand up against odds and think straight rather than an exhausted one who looks for rest, which enables him to survive.

A word here for young Company Commanders. It is often seen that a young officer who abounds in physical energy and fitness, presumes the level of fitness of his Company is commensurate with his own. We as Company Commanders should realize that all five fingers of the hand are not of the same size and specifically in relation to the company, not of the same age either. A young soldier and a not-so-young soldier cannot be driven physically with the same force. Instead of building them up physically, one may cause serious injury.

Now a suggestion for competitive Company Commanders. Whereas competitions are an ideal method of extracting the best out of a man, the aim of winning should not be achieved at the cost of unjustified torture exercise of the company, which may prove to be counterproductive in the long run.

Last but not the least preparation should enhance the specific capability of the soldier for a particular terrain. If the area of operations is mountainous then exercise and physical conditioning should aim at building the body accordingly.

(e) Modification of Equipment

The equipment, weapons and vehicles authorized to an infantry battalion for conventional warfare are as per the WET (War Equipment Table). Some modifications in the scale have been catered for, keeping in mind the varied terrains available in the country. However no modifications or scales have been catered for LIC, despite the fact that it has practically become a permanent fixture on the priority list of the army.

Experience in LIC over the years has time and again exposed the limitations of our vehicles, weapons and equipment. There are a few amongst the planners who have addressed this basic need of a soldier but resources and priorities have at times outmanoeuvred their noble designs.

Having faced the brunt of the lack of suitable equipment in LIC for approximately six years, my unit has carried out some modifications at its level which proved to be fairly successful and suitable. Most of these modifications are also being carried out by units presently deployed in LIC and have stood the test of time. Some of the unit-level modifications are:

- Modifications of vehicles was the single most important improvisation carried out by the army since OP PAWAN. We have continued to use this over the years and have found it to be reasonably effective. The rolled up tarpaulins have been able to give the soldier a fighting chance against an ambush or even a sniping insurgent. The placing of metal sheets or thick wooden planks give some safety from angular hits and deflected bullets. In the IED threatened areas, sandbags on the floor of the vehicle give some protection, unless the insurgents plan on the drum of explosives like the LTTE did during OP PAWAN. The camouflage nets also help in stopping the lobbing of grenades in urban areas like Srinagar, Anantnag, Doda, Sopore, etc. The bracket for mounting automatic weapons on the hood of the vehicle also proved to be very effective as it provided the vehicle a 360 degree area of action.
- A basic modification carried out on the 7.62 mm SLR was to adjust the position of the rifle sling which was proving to be more of a hindrance rather than an aid for the carriage of the weapon. Keeping in view the requirement of carrying the weapon in 'ready' position at the general level of the waist rather than slinging it on one shoulder or carry it in 'shikari' position, this modification was carried out. The simple modification requires the front 'sling kari' to be opened and fixed ahead of the slot or reversed depending on the convenience of the user, and the rear end of the sling is tied to a strap which is fixed around the small of the butt. This enables the rifle to be held straight on one

or two shoulders in a perfect firing position ready for action.

- Based on the principle of Infra Red beam-directed firing, wherein the weapon is able to fire on the target which is illuminated by the IR beam, a torch is strapped on the forehand guard of the rifle and its wires connected to the trigger. As soon as the trigger is pressed for the first free motion before the actual firing action, the torch light will switch on, illuminating the target till approximately 25-50 metres depending on the quality and power of the torch. A point to be borne in mind is to ensure that the torch is adjusted to produce as concentrated a beam as possible. This will increase the accuracy of fire. The modification was very useful in bringing down accurate fire on targets from the hip position in snap-fire situations. However, this modification can only be used by responsible personnel because of the danger of giving away of one's own location.
- Light is also used in another modification to get accurate fire on a target. Here the basic requirement is that the target should be illuminated or clearly visible. The simple modification requires a small bulb to be tied close to the foresight so as to illuminate it for aimed fire. The principle for switching on the light remains the same. Luminous paint should also be put on the foresight to enhance the effect of light. This was quite helpful in getting down aimed fire at night. A single 1.5 volt cell was adequate to achieve this illumination.
- The Pack 08 and the rest of the web equipment have always proved a cumbersome load-carrying support system. Experience proved that the majority of the engagements did not last more than 24 hours including the time of journey to and fro from the target area and the time for the engagement itself. However the equipment though designed to cater for such operations has proved to be a deadweight rather than a weight-carrying aid. With an aim to make carriage of the basic loads for a 24 hour operation including on-weapon ammunition, grenades, rain cape/OG (Olive Green) jersey, food, map, etc. a battle vest was designed. This is shaped like

a modern-day vest for babies to help keep their dress under it clean while eating. The vest which is open on both sides is tied with the help of straps at the waist and chest level. Its broad shoulder straps help to distribute the load evenly. The front pockets are used to hold magazines, grenades and miscellaneous items like notebooks, chinagraph pencils and service protractor. The rear pouches hold the rain cape, map and the food. The vest is made from tarpaulin or thick cloth with a camouflaged cover to make it presentable. Two straps at the rear bottom of the vest fasten the water bottle and make the battle vest a complete equipment carrying workhorse in our present day LIC scenario.

— Batteries for our radio sets have always proved to be a stumbling block for communication requirements. The shortage is a bitter truth we all live with. The answer to this was found by my radio operator Naik Ajay Kumar. He opened four to five batteries at one time for our ANPRC-25 and used the age-old principle of recharging the dry batteries by keeping them in the open under the sun. By this method I was able to get five days usage instead of a recommended output of one day and at best a miserly output of two days. We also used the 1.5 volt batteries inside the ANPRC-25 to activate our night vision devices on the post and Motorola radio sets which had been issued as sector stores.

— Latrines I am sure must be a nightmare for most of us and our command in the field areas and LIC. The age-old system of STLs, DTLs (Shallow/Deep Trench Latrines) and the 'thunder box' have more than outlived their utility. After getting our first success in Doda, the company received a cash reward from the CO. I asked the company during an informal get-together as to what would they like to purchase. I got the usual suggestions like steel crockery, steel thermos flask, new curtain cloth etc. I had already made up my mind about the purchase. My only task was to present it to the company in a suitable manner. I gave them the circumstances and difficulties under which they go to the latri Des, at times under the effect of rains. The vivid and graphic description was enough for them to readily agree to my suggestion of going in for eight fibreglass latrine seats which have since proved very hygienic and effective for my company. I know that it is no place for them to sit and remember me, but alas, so be it!

(f) Deficiency Management

The warning orders for move have always been accompanied with the realization of major deficiencies of manpower and equipment. These deficiencies look routine in peacetime functioning, but their absence can prove to be life-saving factors in LIC. It is absolutely essential to take stock of the holding of a company immediately on receipt of the warning order. The areas which will require an additional look while raising these deficiencies are:

- Automatic weapons, specially MMGs (Medium Machine Guns).
- 60 mm mortar and 84 mm RL (Rocket launcher).
- First Line ammunition for the weapons mentioned
- Night vision devices.
- Radio sets, GPA (Ground Plan Antenna) and telephones.
- Tentage.
- Engines for class IV vehicles.
- R-l (i.e. First echelon repairs) spares.
- Clothing.
- Stationery.
- Oil cookers and burners.

At times it will be prudent to carry out liaison yourself to get some of these stores released because the staff officers will perhaps remain "too busy" to do so. Keeping in mind the terrain and weather conditions, it will also be sensible to take up a statement of case for the issue of jungle boots, which have invaluable use in Eastern LIC areas.

(g) Representation of Supporting Arms

LIC is generally accepted as the battleground for the foot soldier—the infantry man. Without support and cooperation from the supporting arms the much-needed specialization can severely undermine the efforts of the army.

It has been seen that representatives of the supporting arms, marry up with the infantry in the area of operations itself. However this precludes the important aspect of combined training prior to the move. The building up of a bond of mutual respect and healthy working relationship require a training area where uninhibited study of inter-arm functioning can be carried out. This may not be possible in a tension-filled scenario and may well become the cause of serious misunderstandings.

The joint training programmes also help in ironing out basic disputes of 'who will do what' which crop up as avoidable incidents on the post. Amongst other aspects, going for operations and giving duty on the post under special circumstances are some of these.

The training also helps the service personnel in understanding the problems, limitations and characteristics of the equipment which helps them avoid making unreasonable and impractical demands on one another.

The aspect of physical conditioning which has already been discussed in detail, is as important for the accompanying representations of supporting arms as it is for the fighting troops. There have been numerous incidents of troops of supporting arms lagging behind the others. This is not only detrimental for the cohesive action of a group but it also endangers the lives of the men.

(h) Screening of Counter-Insurgency Films

We are all very well aware of the power of television. The combined effect of sound and motion pictures produce a lasting impression on the mind of the audience.

This should be made use of to explain the specialized operations of LIC. Suitable modification to these can always be made keeping in mind the terrain available and unit-level drills.

A number of useful films are available with the Counter Insurgency and Jungle Warfare School and provision of recording them on payment exists with the School.

Notes

1. Lt Gen KS Brar, *Operation Blue Star*.
2. White Terry, *Swords of Lightning*.
3. *Gwari* is a summer house on the meadows where people stay and graze their cattle.
4. Goldhamer, Herbert, *The Soviet Soldier*, p. 104.
5. *German Army Paper*, 1971.

Arrival

"It was 1700 hours, and time for the evening transmission oft he HM cadres. The crackle of the radio set broke the silence of the beautiful evening's serene, quiet atmosphere. The birds perhaps were the only other constantly chirping companions to the section commander code-named 'Two'. He looked worried as he got ready to give the report of the newly-inducted battalion. Within five minutes the HM cadre sitting at the vantage point near Puloi had passed the information to his commander at Kehlotran. Orders have been passed to move higher into the mountains and to carefully watch the movements of the newly-inducted battalion. A local political leader was instructed to go to the camp and find out as much as possible about the 'dogs'— the Army."[1]

(a) Reaction of the People

The all-too-familiar, thronging crowds lined up to welcome an army — the saviours, is not what the usual response of the people is on the army's arrival in the LIC environment. From slogans like *"Innocent People-Killing Force"*, a corrupted full form for the acronym "IPKF" (Indian Peacekeeping Force) painted on the walls at the Madras Railway Station before induction into OP PAWAN, to cold steely penetrating, at times hatred-filled glances in Srinagar —the signs are all there for the soldier to comprehend and realize that he is by no means a welcome visitor.

In the first two to three days after the arrival of your company in the area of operations, the attitude of the local people towards the government and the army will become quite evident. In most cases the reaction will be that of neglect and apathy rather than antipathy. Very rarely will anyone like to willingly tread on the toes of the army and check if any reaction is forthcoming. However areas having a bi or multi-responses polar ethnic or religious hatred and relief. This kind of situation was very much evident in OP PAWAN from the responses of the Tamils and the Muslims; in

Assam, from the Hindu Assamese and the Muslims and in Doda from the Muslims and the Hindus.

The reaction obviously cannot be a thumb rule for all areas hit by LICs. There will certainly be pro and neutral villages and pro and neutral pockets in towns. This is the reason why it becomes important for the Company Commander to assess his area. This will help him to decide upon the attitude and strategy in the ensuing days.

The initial days can be compared with the first date with a girl. The attempt to impress and create a good impression is made by both the boy and the girl. Hesitation, caution and excitement are the pillars of this association.

During these initial days, you will also find some local 'elders' or leaders taking the initiative to meet you. Beware of them until you can establish their credentials and see through their intentions. For all you know, they will be the very moles sent in by the insurgents to get first-hand information about you in particular and the organization of the post in general. It is advisable to keep them at arms-length distance from the post and meet them at an ad hoc visitors' enclosure.

These visits can also serve your purpose. More often than not, any comments or reaction expressed will find its way to the insurgents. Therefore these 'leaders' are an ideal channel to send a word of caution and a signal for a negotiated surrender to the insurgents. Surrenders have always been the easiest to get either when a unit has initially arrived, or once they have established themselves as just and humane. However desirable, the latter may take years or may not happen at all.

(b) Protection Arrangements

Another factor which merits attention is the provision of rudimentary but effective protection arrangements. The initial days are characterized by an extremely careful and at times opportunistic insurgent and careless and overconfident soldier. Incidents of sniping and jitter actions have been quite common in the first few days to 'check out' on the drills, procedures and level of preparedness of the security forces. It will be advisable to have strict control on the movements and actions of the company. The sentry posts should preferably be supervised at any given time of the day by a duty JCO.

Latrines may not be dug on the first day. It is again important to ensure that the troops do not move away from the prescribed areas. An armed escort is not only desirable but essential.

Sentries should also be placed on vehicles and stores during the process of unloading and settling down. These are lucrative targets which do not even require insurgents themselves to destroy. Innocent looking civilians are enough to cause extensive damage.

Strict control should be maintained over access to, and in the vicinity of the post. No sneakers and casual onlookers should be allowed. Helping hands in case required should be checked and then allowed inside the camp.

Perimeter patrols around the location of the camp in a radius of approximately one to one-and-a-half kms are recommended to ensure that no misadventures are undertaken by the insurgents. This will also create a sense of security or caution, as the case may be, among the people.

(c) Logistic Support

Logistic management becomes a very important factor requiring immediate attention during the initial days. A couple of days packed food is not adequate to suffice the needs of the Company. A well-planned and coordinated set-up is desirable and specially so in areas which are inaccessible to vehicular movement.

Whenever it is not possible to carry the stores of the Company in one trip, utensil sets should be carried within subunits with the bare minimum essential requirement of kerosene oil. In case the establishment of the post and carriage of stores requires time and effort, local resources can be tapped on a strictly returnable basis.

Once the stores have fetched up, the langar* should be established in a depression to avoid giving away the location of the camp during the early hours of the day to sniping action, by the glow of the burning fire.

The Company should not be allowed to mill around the langar during meal times. This will present easy targets for the opportunistic insurgents eager for a breakthrough.

Before moving into the post the requirement of condiments and reserve rations should be projected. This will depend upon the weather conditions and accessibility of the post from the Battalion HQ. Based on the area of operations the use of civil transport can be made judiciously for selected stores.

MOH (Meat On Hoof) will always prove to be a problem area. The poor health of the available animals gets aggravated by the not-so-comfortable rides in vehicles, at best, and on foot over mountains at times. Viable alternatives will have to be worked out based on availability and local policy.

In mountains, the use of local mules is essential. Keeping in mind the threat perceived, either 'stops' or patrols will have to be sent to ensure uninterrupted move of stores. The documentation for release of money to the porters and ponies is a complex affair and requires meticulous records and details to be maintained. I have known two incidents wherein the unit representative had to

---

* Cookhouse.

go back to the area of operation to make payment to the porters!

(d) Area Familiarization

Familiarization is an ongoing process which starts the day the company reaches the area of operations and it does not get over even after three years of stay at the post. However, the basic requirements of getting to know the place in the correct perspective, and more importantly, knowing the insurgent and his sympathizers' broad strategy is essential. This must start immediately on arrival.

The area should be divided into sub-sectors in concentric circles. The size of the circles depends on the population. However five kilometres is a realistic target to begin with. Each sector in this circular area must be visited by at least all JCOs and section commanders immediately on arrival. During these initial days the patrols should personally be briefed and debriefed by the Company Commander to ensure that a foothold is established as planned. During the debriefing it is suggested that all available NCOs attend, to reinforce the names of places and routes for movement.

During this process of area familiarization detailed and systematic information gathering must continue. This information should be documented and records should be made for future reference. Maps should be updated and names of places found unmarked added. There is a saying that *"Information is knowledge, and knowledge is power."*

In case some areas are inaccessible without guides, locals should be made to accompany the patrols. However, due care is required to be taken in the selection of guides. Ideally, the guide should be picked up at the shortest possible notice to avoid any chances of being set up for the insurgents' ambush.

While on these area familiarization patrols, it is useful to look out for places which can be likely ambush sites or hideouts for the insurgents. These places should be discussed during debriefing and marked on the map for every patrol to see before moving out.

There may be a requirement to stay the night at villages if the duration of the patrol is more than a day. This is a double-edged weapon. Whereas there have been very few instances of army patrols being ambushed during their stay in a local's house, yet this cannot be ruled out. The best bet will be to stay at the house of *a known sympathizer of the insurgents.* This will preclude any kind of attack, jitter action or interference with the party. I will like to specially emphasize that the urge to stay at the house of a sympathizer or source should be strictly avoided and junior leaders should also be dissuaded from doing so.

During these area familiarization trips the ideal course of action is to listen and let the people talk. Experience has shown that the locals of any area give away useful information in their exuberance to impress the officer. Wherever possible, but not at the cost of arousing suspicion, a sensible literate NCO should be deputed to note the contents of the conversation. These can subsequently be used to place together the numerous jigsaw puzzles that present themselves.

While familiarizing with the terrain, it is also essential to familiarize oneself with the people and the insurgents. The basic nature of the people, their attitude towards the army at large and specifically the company will become evident. It is at this stage that a bond of understanding can be initiated, which can subsequently be nurtured. One realizes during the course of operations that it is very essential to understand the psyche of an insurgent to enable correct counter-measures to be taken against him. I would like to describe some facets of the intrinsic nature of an Afghan guerrilla, as an illustration.

Afghanistan over the years has been the battle ground for tribal clashes and bloody wars. This lifestyle has been the training ground for the young boys to be blooded. The simple character of an Afghan can perhaps be summarized as *"a loot-hungry, inflexible, courageous, gun-toting, fanatical guerrilla who, during the course of his difficult life-style can be extraordinarily hospitable yet stubbornly revenge-hungry, and who only, fears being maimed for life."* An incident may well describe the foolhardy method of two guerrillas proving their bravery to each other.

"A small group of Afghans were clustered around a fire, arguing. Two of them were arguing as to who was the bravest. To prove his point, one of them leaned forward and thrust his hand into the fire. He held it there with the flames eating at his flesh. Despite the excruciating pain he made no sound. Only the locked jaw, the screwing up of the eyes, and the slight shaking of his arm, indicated the supreme effort of will necessary to conquer the pain. For a few moments he kept it roasting in front of his audience. When he pulled back his hand it was bright red, dripping blood. The man had established his courage."[2]

Another bewildering quality, or rather a characteristic of the Afghan guerrilla is the fear of getting maimed as the following passage conveys.

"He knew fear, but not the fear of death. If ound that most were a fraid of mines and were hesitant to attack posts closely protected by minefields. Their concern was living the life of a cripple, in a society where physical stamina and hardiness are indispensable. Mines tended to blow off feet: could a man raise his family, tend his sheep, build his house and climb the hill without his legs? The prospect of such a life was infinitely more fightening than death on the battle field."[3]

(e) Extreme Caution

I will like to emphasize the need for extreme caution even at the cost of repetition. It is often the first few days that make or break the will of the army or the insurgents. A little care, hard work and caution may make the difference between being the victor and the vanquished. Death can take months to resettle and start the fight against insurgents.

Notes

1. Contents of a message passed on the arrival of the Rashtriya Rifles in Doda District while relieving the author's battalion.
2. Yousaf Mohammed, *The Bear Trap*, p. 32.
3. Ibid, p. 32.

Establishing a Post

The siting and ultimately the establishment of a post is like building a home, the only difference being that the wrong siting of the post can easily result in loss of men, material and equipment whereas the making of a home may not cause such disastrous consequences. At times it will be prudent to select a tactically suitable location rather than a comfortable and palatial building.

(a) Selecting a Site

The location of a post should, wherever possible, be on a dominating piece of ground. The area should 'overlook' the neighbouring areas rather than be 'overlooked' by the surrounding areas. In other words bowls and depressed areas should be avoided.

The site should not be in a thickly built-up area. This increases the chances of being sniped at from very close quarters. It also gives limited fields of fire and becomes an obvious deterrent to opening fire. However where the requirement of opening the posts in a built-up area is inescapable, school buildings, police posts, Public Works Department godowns, rice mills and abandoned factory premises are suitable sites.

The house or building should not be mud-plastered. In case it is, it will not be able to provide protection against small arms fire. However if no alternative exists the side walls can be reinforced with the help of sandbags or timber sleepers.

The prospective site should be near a road head to facilitate easy movement of stores, supplies and most importantly quick reaction teams (QRTs). In case it is possible it should be on a highway so that movement on the road need not be accompanied by road opening for everyday moves.

The site should have an independent water source within the premises. This is important for ensuring safe drinking water and bathing facilities within the post. In case water is carried through over ground pipelines, immediate action should be initiated to convert also necessary to ensure that the water supply is from the main water reservoir and not from an independent tank. This rules out the chances of water poisoning.

(b) Taking Over of Accommodation

More often than not the accommodation selected for establishing the post will belong to an individual rather than the government. It is essential that basic completed to ensure that whoever helps the army does not end up being a loser.

In case of government accommodation, the only requirement is to inform the highest civil official of the area and request for the accommodation in writing. My experience has been that respectful and humane behavior never fails.

However in case of an individual's house, after obtaining his willingness, it is suggested that the following actions be completed to enable him to get rent for the house:

- A sketch of the house should be made with the dimensions of all the rooms.
- A letter should be sent to the civil administration along with the sketch for approval.
- On obtaining approval, insist on the representative of the administration to come to the post and inspect the house for

the correctness of the sketch and for fixing the rent.

– Pursue the matter and ensure that rent is given to the owner of the house at the earliest.

It is also important to ensure that the owner of the house is accommodated in another house in the same village, if he so wants. Here the good offices of the village headman a n d coercion on your part will stand in good stead. I will like to relate my experience in Doda, which enabled me to select a tactically well-located house in the area.

On my arrival at the village where I had been tasked to establish a post, I short-listed three houses which were tactically suitable and met my requirement of accommodating the Company. I called for the elders of four closely-located villages to discuss our stay in the area. Meanwhile, I did a thorough recce* of the three houses and evaluated their suitability keeping in view my requirements. I also discussed this with my Company Officer and JCOs. We all came to the conclusion that the house at the top was both ideally located and had the least number of people requiring displacement.

The other two had either a big family, or, only women staying in them. Having made up my mind, I sat sipping "kahwa" and patiently waited for the village elders who slowly trickled in after their '*jumma*'[2] prayers. After briefly introducing myself I said, "Keeping in view your long-outstanding demand for an army post in the area, the government has kindly agreed to it." I waited to see the effect. Not very many amongst the audience seemed happy to hear the news. I continued, "I am sure that you will go out of your way to help the army achieve its aim. I have been sent here to serve you and assist you and the civil government in crushing insurgency." A twitch of the lips and a wince were more than evident on the otherwise expressionless

---

* Reconnaissance

faces. "I will be requiring your support and blessings to achieve this aim. But right now, I only require a small place to open the post. I know that you all are famous for your hospitality and would like to offer the most luxurious house to me. But my requirements are very meagre. Out of the three available houses that are suitable for opening our post, I would feel ashamed to ask for two of them as the one on my left houses women, and the other one has four families residing in it. I am sure that you will agree with me that the modest house on top is adequate for the post," I continued. Before the owner could give the tactical errors in selecting his house, I said, "I am also sure that you people will kindly decide amongst yourselves as to where the owner of the house can be shifted." Within one hour and forty-five minutes, my company had shifted into the house, thanks to the cooperative attitude of the villagers. I did ensure that the owner of the house received rent for the house as soon as was possible, and also that he got a comfortable house to stay in within the same village.

Tact and respect for the sentiments of the people had helped me select the house that was ideally suited for my post.

(c) Defence Capability : Analysis and Layout

Though the selection of a post involves keeping in mind the defensive capability, a detailed analysis may not always be possible. This should be done at the earliest to ensure that any contingency can be dealt with effectively and efficiently.

It is advisable to carry out this analysis by going to a distance all around the post and visualizing the places from where the post can be engaged. Having selected these places, their distances and back-hearings should be taken. After this, planning should be carried out in consultation with the officer in case present and the JCOs. While placing the weapons, it should be ensured that the bunkers have adequate place for five to six personnel to fire from. Ideally arrangements should be made to get to the bunkers without being exposed to insurgents' fire.

It may also be necessary to place an observation post at a height nearby to enhance the observation of the area.

During this process it is essential to find out the ranges of target areas around. This may not be possible by fire in built, up areas but in mountains it can be done by speculative fire of MMGs with tracers. This fire will also create an impact on the people which will not only build confidence but will dissuade any misadventures.

Having completed the analysis, the task of construction of bunkers will be undertaken. Depending upon the local material available, this can be accomplished with the help of the villagers. Stones and logs of wood are good substitutes for sand bags which generally remain in short supply.

While constructing the bunkers the rear side generally remains exposed. Keeping in view the chances of fire from made that direction, adequate protection arrangements should be made.

(d) Security Lights

The ideal condition for a post is when the sentries can see the area around the post, but an outsider cannot see them. This can be achieved by the method of security lights. These should be placed at approximately 25 to 30 metres from the post facing outwards at prefixed angles lighting the area ahead. The lights should be connected in parallel, and not in series, to ensure that even if one light fails the others keep functioning.

I am reminded of the STF (Special Task Force) post at Vellore colony Batticaloa (Sri Lanka). The company post besides having a self-sufficient complex had near-impregnable underground dugouts for the sentries, towers for the OPs (Observation Posts) and an excellent system of security lights which had made movement up to 200 metres around the post nearly impossible at night. The post could not be sniped at, either during the day or at night.

(e) Barriers

In case the post is located on the road, it is essential to establish barriers 50 metres on either side of the post, to force all passing vehicles to pass through them at a slow speed due to the zig-zag movement. This precludes any attempts of hit-and-run tactics. The approximate distance of 100 metres gives adequate opportunity to the sentries to get the insurgent.

The barriers also function as vehicle check posts by adding requisite manpower and boards requesting the vehicles to stop. This creates a feeling of caution among the insurgents and has a good deterrent effect.

If metal structures are not available for construction, the barriers, drums or even tree-trunks serve the purpose. The aim is to make the vehicles pass through the barriers and not over them. Adequate arrangements for lights should be made to ensure that no accidents take place at night.

(f) Nets

In built-up areas grenades are an entirely avoidable nuisance. Lobbing of grenades by women, children and old men is quite common and has to be taken with the seriousness it deserves. This is all the more relevant where the posts are located in built-up areas and along major roads. A simple yet effective method of avoiding casualties to troops is to put camouflage nets on the exposed portion of the sentry post.

(g) Arrangements for Guards

Imagine a situation where the sentry detects hostile movement of insurgents close to the post. Of the two sentries posted, one is forced to either shout, whistle or go and ginger up the others to elicit any worthwhile response. It is very important, and specially so in an extremely hostile environment, that the complete guard for the night sleeps at the sentry post. This strength of eight other ranks is quite capable of dealing confidently with most contingencies without having to leave their post.

(h) Alarm

The first thought that comes to the mind when one hears the word "alarm", is either the howl for a thief, or the shout for fire along with the continuous ringing of the quarter guard bell. However in LICs it may not always be desirable to raise such an alarm. It will not be sensible to shout and ensure that the prying intruder vanishes into the night. An alarm system should be able to get the company ready for action at their prescribed position, and yet give an impression to the insurgents that the post has not noticed their movement.

According to my recommended system, no sentry is required to leave his post, and yet is able to caution the company. A wire is connected from all the sentry posts on to a switchboard with small bulbs. Numbers of the sentry post are written under them. As soon as any suspicious movement is seen the sentry presses the switch at the post, which in turn lights up the bulb on the board (preferably with the operator). It is then just a matter of a well-rehearsed drill for the Company to move to their designated locations.

(i) Quick Reaction Teams

This is a much discussed concept. QRTs are often seen to be not as quick as is desirable. Laxity creeps in over a period of time when their usage is negligible and practices have been done away with.

I will like to emphasize that one might as well not react rather than reacting late. A futile action will not only project the Company in poor light to the people, but will also affect the morale of the Company.

Regular practices of moving the QRT should be carried out. These should be timed and the party should be checked for all the stores required to be carried.

The composition of the QRT should be clearly spelt out and personnel should be nominated by name. The team should sleep together and not be put on guard for the night. It is also important to have a balanced party. Mere numbers should not be the criterion of nomination. The Company Commander may like to nominate certain individuals who are capable of independent actions themselves.

(j) Vehicle Park

The vehicles should preferably be parked away from the road and out of the reach of nosy commuters to avoid any chances of damage from sabotage. The ideal situation however, is an enclosed space for vehicles giving them safety and providing easy maintenance.

(k) Interrogation Room

Interrogation requires a detached room, where questioning can proceed without the interference and pressures from local people. The room should be as far as possible from the visitors' room or shed. It is also desirable that movement of the suspects from the interrogation room to any other room in the post is not carried out from anywhere close to the visitors' room.

(l) Visitors' Room

This should be away from the main post. A sentry should be placed here to regulate the movement of the visitors. Depending on the work required to be carried out, the visitor can be brought to the main complex by the sentry. It is desirable to place a water container and some glasses in the visitors' room.

A thorough check of all visitors should be carried out before they are allowed inside the post. This practice should continue unless the credentials of the particular visitor have been established beyond doubt.

Old people and visitors coming from distant areas should be offered tea. This establishes a feeling of confidence and respect in the minds of the people which is absolutely essential for gaining information and breaking the barriers of hesitation.

(m) Axis of Maintenance

After moving to the post, ensuring that maintenance is carried out without any interference from the insurgents becomes imperative. Depending on the terrain, this may have to be ensured by protective arrangements like helipad protection, ROPs (Road Opening Patrols) or placing stops.

It may also be possible at a number of places to utilize civil transport for the movement of supplies. However, this decision requires the Company Commander to weigh the situation and intentions of the parties involved before taking this step.

Stocking of supplies can reduce the problem of frequent movement and can be resorted to. It may become necessary during the monsoons.

(n) Consensus

Lastly, I strongly feel that all decisions while establishing posts should follow the course of consensus. While the ultimate decision is the Company Commander's, LICs present numerous situations which can only be resolved by decisions which are neither prescribed nor taught by anyone. Here ideas can come from the least expected sources from within the Company.

The opportunity to discuss also pushes the junior leaders to think about LIC, an act which is required to be done continuously, if survival of one's command is to be ensured. This also gives the commanders at lower levels the satisfaction of being a part of the decision-making process.

Notes

1. A country cousin of tea made from special tea leaves and consisting of number of ingredients. A speciality of Kashmir.
2. Urdu word for 'Friday'.

First Impact

Here is the experience of Lt James McDonough, a platoon leader of US forces in Vietnam, with a private in his platoon who having had his stock of drugs destroyed by the Lt, wanted to convey his displeasure.

King was carrying M-79 grenade launcher as his weapon. I checked him out and moved up the line to inspect the other soldiers. Suddenly there was a hollow-sounding explosion and a whistling past my ear, followed by another explosion a second later about one hundred and fifty meters out in the bushes. I spun around to see King with a sneer on his face.

"Sorry, Lt," the words hissed from his mouth. "I dropped my weapon."

The grenade launcher dangled from his right hand, a waft of smoke curling up from the barrel. It was clear he had fired the 40 millimeter round at me. At that range he could not have missed so he must have meant it as a warning. He was telling me he did not like my attitude.

I was scared but I was angry too. King had not only threatened my life but my authority as well. A few weeks earlier I might have been overcome by the situation, but since then I had developed the thick skin necessary for survival in Vietnam. With murder in my eyes I looked into King's face. Slowly, deliberately, I walked up to him, saying nothing until I was close enough to touch him. I had made no threatening gesture. Suddenly when I was close enough to smell his breath, I clicked my M-16 to the fire position and placed the muzzle of the weapon in the cavity of the underside of his chin. He went up to his toes as I increased the pressure of the rifle building up against the bottom of his head.

The sneer was now on my face as I raised my voice, so everyone could hear, "You son of a bitch." The eyes of everyone in the platoon were riveted on us. "Drop your weapon again, and you'd better pray to God it hits me square in the back. Because if it doesn't, you bastard, I'll blow your brain to kingdom come!"

Not a man stirred. My finger touched the trigger gingerly and King's eyes widened. The sullen look was gone from his face and beads of sweat formed on his forehead. Slowly I rotated him as he stretched to keep his toes touching the ground. He, as well as the others, knew that the slightest twitch would send a spray of bullets up through his face and out his head. For thirty seconds I let him hang suspended in terror. My rage had passed, but I was committed to play the scene out to its conclusion. Then, slowly, I lowered him off the barrel and flicked the weapon back to safe.

"All right, let's go." I tried to sound calm.

As I walked away I awaited the blast that would tear my head off. I was betting that his bravado was only a bluff, that he really didn't have the nerve to kill me, but the choice was his. To turn back, to watch him, would ruin my act. Now I was the one who was sweating. Butt could not flinch, and 1 did not. I gave the order to move out, and we left. King took his position in fine and went through the wire with us. I had won.

(a) First Impression

The first impact, the first impression always wins or loses the battle for supremacy. This is specially so in LIC where it is almost as important to create a good impression as it is to be good. The impression created is like the review of a book or a film which makes others interested or disinterested in 'buying' the story.

This by no means implies that we go out of our way to please the people. But it also does not imply that we go out of our way to shout them away as sympathizers of the insurgents (which they may be). Luckily for us, the public at large still feels that the army is an organization which does what it is told to do and has no malaise or hatred for any class or community.

(b) Speak to Elders

It is very important to convey the basic system of functioning to the people. Every Commander has his style under the general guidelines issued by the Battalion HQ. There are a number of issues on which the people will like to seek your views. It may not be possible to speak to all the people immediately, but it is certainly possible to speak to the village elders.

The congregation should comprise all village and block level political workers, 'nambardars' (headmen) or 'gaon burahs', (village elders) chowkidars, representatives of all religious groups, headmasters, senior teachers, and any other village elders who may like to come.

The best method of talking to them that I tried, and with success, was to request their spokesman to give his views and also ask all those present to give their opinion about insurgency in the area, the role of the civil administration, the role of the army, the help that the people were willing to give to help solve the problem of insurgency and any developmental projects that they would like the army to project for their area.

The results of this unique get-together were astonishing. The rivalries in terms of religion, politics, village level disputes and even different militant groups came up in the open, with all the speakers trying to put down the others. The power equations of the area became quite clear. In their exuberance all of them promised help to the army and condemned insurgency in no uncertain terms. In fact I was quite impressed with the rhetoric in some cases. A lot of them would have made good orators!

On my part I gave them broad guidelines regarding my way of functioning to include:

- No punitive action will be taken against relatives and family members of insurgents in our areas just because of their relations.
- Surrenders with arms are most welcome, because we had come to finish insurgency and not the insurgents, if they cooperate.
- The army has come to ensure their safety, but it also expects their cooperation.
- The days of insurgency in the area are numbered as people at large had seen through the false Pakistan inspired propaganda.
- The post is always open for people who are willing to help, but would prove to be a nightmare for those who deliberately choose to hoodwink.
- No punitive action will be taken against people who have unwillingly been host to the insurgents before the army came (benefit of doubt given to all).
- Developmental work will soon be initiated with the help of the civil administration.
- The post is very thankful to the people for their help in its establishment.

The results of the meeting were evident. The ice had been broken. The cold and apprehensive looks had melted though not completely. The next meeting with all residents had also been fixed for the next 'Jumma' day.

(c) Meet All Residents

There is a subtle difference in meeting a responsible elderly lot of people and the public at large. The public is not always swayed by sound logic and reasoning but by issues which are easily understood by them and directly affect their lifestyle. This is the guideline that should be kept in mind while speaking to the public at any congregation.

The points that can be conveyed to the people are:

- Welcome all the people and specially the ones who come from distant locations.
- Share the feeling of suffering of the people who have been troubled by the insurgents for so many years (everyone loves to feel sorry for one's self).
- While denouncing insurgency give a graphic account of Pakistan's selfish motives in instigating the insurgents.
- Welcome all misguided youth who want to surrender and assure them help and support.
- Tell the people that they are the 'eyes and ears' of the army and their cooperation will help the army destroy the evil designs of glorified thieves.
- Explain to the people how development has suffered ever since insurgency started.
- In case of Kashmir, tell the people that the Afghan terrorists are taking advantage of the simplicity of the people and are misbehaving with the girls in the area. Emphasize that most of them are infected with AIDS and will destroy the very fabric of society.
- Explain to the people that insurgents are like fish who cannot survive without water, i.e. the local people.
- Tell the people that the army has decided to start developmental work with the help of the people. Give the list of projects approved or which await approval. However one finds that a number of projects can be started at the village level itself. One such project was rebuilding of schools in Doda which had been burnt down by the insurgents. (Give dates of recruitment if announced.)

— Lastly warn the people that you already know about the ardent supporters of insurgency amongst the people who are present. Tell them that you are giving them time to improve and help the army.

After the meeting, all the people should be met at their residences in a progressive manner. The talk may be generic in nature but it will help elicit response from supporters and sources without raising any suspicion.

(d) Action

The next few days after the meetings are very crucial. Since the initiative would have been taken and a challenge given to the insurgents, the eyes of the people will be riveted on any action by the army or against the army. This situation should be explained to the company and also the importance of achieving a quick breakthrough. The level of alertness should be high and motivation at the highest.

The conduct of the action can be in terms of recovery of a cache, because at this stage it is the fence-sitters one has to aim for. To be able to switch them over to the army, it is important to prove to them that the insurgents are on the run and successes for the army have started and are there to continue.

(e) Catch and Release

It may not be possible to achieve a major success in the initial days, despite all-out efforts. The next best option is to catch lower level cadres and sympathizers and hold them for a day. The catches should be of innocuous people in terms of active support. This will draw most village elders and local people who will request for forgiveness on their behalf. This is the time when the release should be orchestrated, but after assurances of good behaviour from the elders. Surety should also be asked for from them in terms of certificates in writing.

In case the requirement is there, the release can also be ordered on the 'special request' of a source who would like to establish his credibility among the people.

These acts will help the Company Commander to establish himself as a kind-hearted man, who is willing to release known supporters in return for the help and support of the people.

It will also help in sending signals to the lower level cadres that surrender is a viable option for the insurgents who being sick of the vagaries of insurgency ultimately want to settle down to a peaceful life style.

(f) Participation in Festivals

All classes of our society, be it any religion, caste or tribe get a feeling of satisfaction, when their customs, traditions and festivals are respected and more so when others participate in them. Besides being an exercise in public relations it also helps the Company Commander learn about the people which stands in good stead in the long run.

History has taught us that conversions of public opinion of any kind have only been achieved through wilful and eager participation on the part of the people at large. But this requires the motivator to adopt the lifestyle and basic habits of the people. The example of Mother Teresa proves the theory. After all insurgency is a battle for the hearts and minds rather than a direct military confrontation with the insurgents.

Note

1. McDounough, Lt Col James, *Platoon Leader,* pp. 74-75.

Interaction with Paramilitary Forces

Sadly the interaction of the army with paramilitary forces at the Company level has not been very fruitful. Aspects like unhealthy competition in terms of catches, charges and countercharges of 'source theft', lack of faith and credibility have dogged what could have been a successful partnership.

The paramilitary forces with whom the army will constantly be interacting in LIC are the BSF and CRPF (Central Reserve Police Force). Of these it would be the BSF which will actually be conducting operations in conjunction with infantry battalions.

Before forming a biased opinion about paramilitary forces it would be justified to take into account their limitations in specialized training, weapons and equipment, regimentation and cohesion. But despite all these factors the paramilitary forces have done a commendable job and specially so in Srinagar town.

While it is important to know the limitations of a service, it would be foolish not to exploit its strength.

(a) Meet Commanders at All Levels

After reaching the area of responsibility, the Company Commander must visit the posts of the paramilitary forces to meet the post commanders. A point to note is that irrespective of the level of command liaison must be carried out, because more often than not it is not the seniority but the will to cooperate that achieves desirable results. The meeting should be in an atmosphere of warmth and mutual respect. The points that merit discussion can be:

- Classification of command structure of the force.
- Modus operandi in case the sub-unit is under the command of the Company.
- Assessment of situation.
- Radio frequencies and method of communication.
- Tying up of movement to avoid clashes.
- Feasibility of joint operations.
- Suggestions and Recommendations.

(b) Mutual Respect

Experience has shown that we do not treat the officers of paramilitary forces with due respect and customary warmth expected from the army. The unnecessary and unwarranted superiority complex on the part of overconfident Company Commanders harms the interests of the organization: If one step is taken forward by the Company Commanders in inter-services cooperation, the officers of the paramilitary are willing to come two steps to help and share.

It must be borne in mind that even before we start judging the professional capability of an individual it is much more important to judge the attitude, spirit and will to achieve the aim. The relationship is more likely to thrive based on personal bonds of mutual respect rather than on records of professional achievements.

(c) Gauge Effectiveness

Once personal goodwill has been established, the Company Commander must gauge the effectiveness of the sub-unit. This is not to judge and pass a verdict but to enable the Company Commander to plan his future joint operations. It will directly influence the level of participation, the roles suited for employment and in fact the very basis of joint operations.

The effectiveness of the sub-unit will be amply clear if the following are considered:

- Effectiveness of the sub-unit commander, i.e. command and control, confidence and professional competence.
- State of morale of the sub-unit.
- Results achieved by the sub-unit.
- Reputation in the area.
- Sincerity of purpose.

(d) Impression of Faith

In LIC it may not always be prudent to call a spade a spade. Though it will be very rare, but in case it is found that the sub-unit is not up to the mark, it is entirely avoidable to go out of the way to prove it to the sub-unit commander. Besides cooperation in operations, there are a number of ways in which mutual help can be rendered. It is said never to concentrate on the weaknesses, but on strengths, if friendship is to be sustained. This should be the guiding spirit for a sustained mutually beneficial relationship. The areas where cooperation can be sought besides operational aspects are:

- The use of radio sets. The paramilitary forces have excellent sets which are relatively small and easy to carry.
- The posts generally have a road worthy fleet which can be used for emergency actions and large-scale moves.
- Patrols from the posts can be used in deceptive action in areas other than the projected sites for operations.

Interaction with Civil Administration

At one time or the other we all have come in contact with the civil administration, either during the conduct of aid to civil authorities or specifically in LIC. A vast majority out of us must have had a tiff either because of our varied styles of functioning, or merely because of the frivolous issues of protocol.

In LIC, where the insurgent organisations are becoming more and more unified in their functioning with civil and military wings under an efficient command structure, our divide is still a major pitfall despite numerous efforts at unification of command and a combined services structure. Though efforts have been made to establish a combined services control against insurgent movements in states like Jammu & Kashmir these organizations have been restricted to the top echelons of the respective services. Because of this the higher directions for fighting insurgency have become coherent but at the Company Commander's level and the corresponding administration at the tehsil, this unification in command and a cohesive action has been lacking. In fact at this level a cohesive action has perhaps not been considered at all.

When we talk of winning the hearts and minds of the people, the process cannot be initiated and executed by the army in isolation. India is a vibrant and stable democracy. The people have been brought up on liberal and open-minded ideals of democracy. And having lived in such an atmosphere, they have an intrinsic resistance to curbs on movements, rights and freedom. The army on its part in the course of its duties also gets the powers to deal with the situation in a strict and ruthless manner which is not often appreciated by the people. It is here that the softer approach of the civil administration comes into effect in order to explain and justify the actions of the army. The amalgamation also helps in understanding the capabilities, weaknesses, limitations and functioning of the various services, which is very important for combined functioning.

(a) Meet Administrative Staff

After reaching the operational area it is important to meet the administrative staff. Generally at the Company level it will be the tehsildar in the tehsil who will be interacting with the Company Commander. Though the appointment of the tehsildar is a Class Two post, it will be improper to send a JCO and call him to the Company post. This exercise in public relations will not only establish the Company Commander as a humble and courteous officer, it will also elicit a more than encouraging response from the civil administration.

The administrative head will be able to give an accurate picture of the state of the people and the government machinery, though specific information about the insurgents may be generic and based on hearsay. He will also be able to use his influence and help in the following spheres:

- Hiring and taking over of accommodation.
- Provision of medicines and medical facilities.
- Hiring of porters and ponies.
- Acquiring judicial remand.
- Giving information about demography, census, land and revenue.
- Public unrest and agitations.
- Untoward incidents involving the army and the population.

Besides the administrative head it will also be useful to meet the chief medical officer, the executive engineer and the head of the road construction office. These appointments will be of immense help for the establishment of the post, its maintenance and carrying out worthwhile civic action projects.

In case it is found that the administrative set-up is defunct or has leanings towards the insurgents, it will be ideal to make full use of their good offices for the smooth functioning of the post and subsequently arresting the culprits once adequate proof of their nefarious activities has been collected.

The police is the other service with which interaction will be very regular. It is essential that a cordial working relationship is established and cooperation is secured. While meeting the senior most police official in the area the points that require discussion are:

- Details of police set-up in the area.
- Past experiences of the police in dealing with insurgency.
- List of insurgents and sympathizers available with the police.
- Available photographs of insurgents.
- Procedures followed for handing over insurgents.
- Points to be kept in mind for preparation of an FIR (First Information Report).
- Procedure for taking a police representative for cordon and search operations.

(b) Unified Command

Having discussed the spheres of cooperation, it has become clear that a unified command is absolutely essential at the Company and Tehsil level. Needless to say, representatives of the police and paramilitary forces, and, if available, intelligence agencies will also be incorporated in the set-up.

Before the proposed organization is discussed, let us see how similar organizations are functioning in countries which have experience in combating insurgency for a number of years. The United States, keeping the aspect of overall direction in mind to counter insurgency and subversive actions, approached fighting insurgency by firstly substituting the phrase "counter-insurgency" by "Internal Defence and Development". The functioning of the armed forces in the overall gamut of IDD is called "Stability Operations" to emphasize the destruction of insurgents and provision of stability. This was further broken down into various functions like "Advisory Assistance", i.e. the task of advising and guiding local military commanders, paramilitary forces and irregular forces. The "Civil Affairs" cell was to provide the necessary cooperation between the military and the population. "Populace and Resource Control", "Psychological Operations" and "Intelligence" are the other self-explanatory cells. Let us see the alternatives offered by John McCuen:

Unified planning, centralized control and a single point of responsibility are the very minimum requirements, for a unity of effort which will offer success against unified revolutionary movement... Unity of effort is however extremely difficult to achieve because it represents the fusion of civil and military function to fight battles which have primarily political objectives ... All the political, economic, psychological and military means must be marshalled as weapons under centralized coordination and direction. Unity of effort can be achieved by a single commander as the French advocate. Unity of effort can be achieved by a committee under civilian leadership as the British advocate ... [1]

(c) Joint Action Committee

Having seen the general thought process for a unified command structure of various countries, let us tread on the debatable issue of a unified command structure which is suggested at the Company level in an LIC environment.

It is assumed that the Armed Forces (Special Powers) Act is in force and the situation has been handed over to the army. In this case the chairman of the committee will be the Company Commander. He will be assisted in the performance of his duties by consultative members representing their services. They will be the SDM (Sub-Divisional Magistrate)/Tehsildar, Circle Inspector/ SHO (Station House Officer), BSF Deputy Commandant, CRPF Deputy Commandant, representative of the SSB (Special Service Bureau) and representative of the CID (Criminal Investigation Department). The command structure chart of the 'Joint Action Committee' is given is Figure 2.

In case of specific issues pertaining to sections of the civil administration, representatives of these sections can be called and can be in attendance.

The periodicity of meeting of this committee can be decided upon by the chairman and the members, however fifteen days is a suitable time-frame for reviewing the actions which have been initiated.

The JAC will take action as per the overall directions given by the committee at the district level. The actions initiated will be sent to the district committee in the form of minutes of the meeting.

In case of any fundamental difference in the functioning of the committee, representatives of the services can approach their senior officers at the district level to seek clarifications.

Subjects that will merit consideration during meetings at the JAC are:

- Assessment of prevalent level of insurgency.
- Specific areas of influence of the insurgents which require additional stress.
- Joint Operation.

JOINT ACTION COMMITTEE

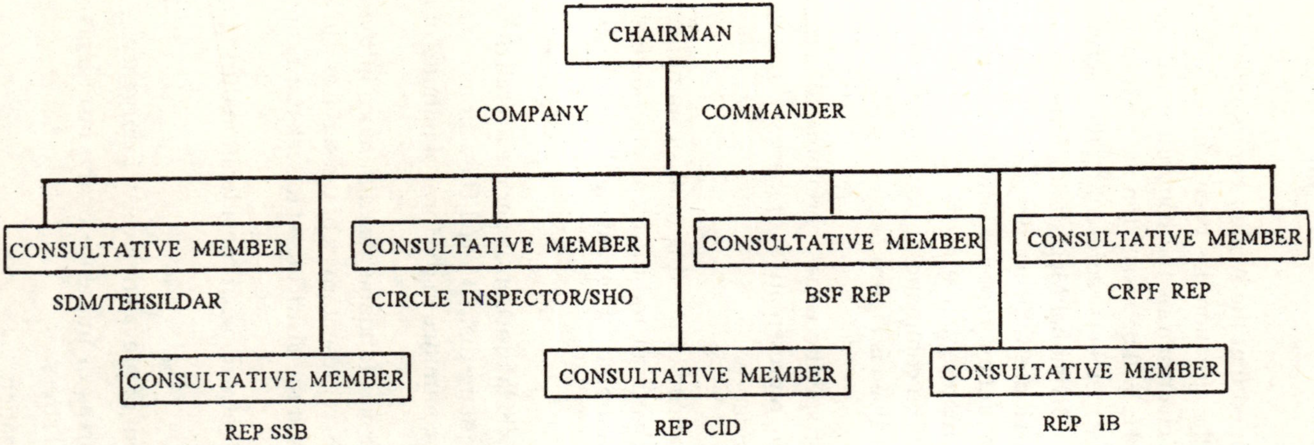

Figure 2

- Intelligence Collection.
- Functioning of the Public Distribution System.
- Functioning of the Public Transport System.
- Functioning of the Financial Institutions.
- Functioning of the Educational Institutes.
- Assessment of action previously initiated.
- Review of Civic Action Projects.
- Discussion of legal provisions for specific cases.
- Assistance being offered by the people to insurgents visa-vis to the government agencies.
- Celebration of forthcoming religious and national festivals.
- Visits and rallies.

One may feel that the points for discussion are secondary to the basic aim of the army, i.e. countering insurgency. Contrary to this belief, the most important job of the army in fighting insurgency is to help create circumstances through which the aims and policies of the nation can be achieved by a popular state government. And for this, mere military operations are not enough to achieve this aim as was adequately proved by the defeat of the French and the US Armies in Vietnam, and the Soviet Army in Afghanistan.

Note

1. McCuen, John, *The Art of Counter Revolutionary War*, pp. 72-73.

Psychological Operations

Most men fight for some goal. When the goal is found unattainable, they may accept a lesser goal. Even when victory is at last accepted as impossible, they may continue; they may still fight on in hopes of obtaining less harsh terms of surrender to allow themselves to be captured, when the material disadvantage of defeat is no longer greater than the disadvantage of continued resistance. That has been the rule for all wars ...[1]

With the passage of time and after gaining practical experience it has been realized the world over that mere anti-insurgent operations are not enough to crush insurgency in the modern LIC environment. Countries which did try to use their military might were embarrassed and forced to retreat by relatively ill-equipped and hostile bands of guerrillas. The experience of America in Vietnam and the USSR in Afghanistan stand testimony to this judgement.

It has been recognized that since 1945, *"of the 160 armed conflicts that have taken place around the world... perhaps three-quarters were of the 'so-called' low intensity variety"*[2]. This being the nature of warfare that will be confronting nations and more so third world nations, it becomes necessary for a country like ours to chalk out its broad strategy to fight the same. Unlike conventional wars, LICs are more often than not protracted in nature. They do give adequate time for formulation of policy, but this often leads to no policy being enforced and further disaster.

Let us see the fighting principle of a Vietnamese fighter which he observed as part of his 'Oath of Honour':

In contact with the people, to follow these three recommendations:

— To respect the people;

— To help the people; and

— *To defend the people ... in order to win their confidence and affection and achieve perfect understanding between the people and the army.*[3]

This forms the basis of fighting insurgency from highest level of war to that of the Company Commander. It will thus be these guidelines that will form the basis of discussion of psychological warfare.

(a) Level of Support

Before commencing on the arduous journey of the study of psychological warfare, it is necessary to judge the level of support that the movement has from the people. It is also important to check out on the particular class of people that have been influenced the most and who form the backbone of insurgency. The psychological operations will have to be targeted towards them. For example while fighting insurgency in Assam against the ULFA, the target groups were the students and the intelligentsia, unlike Punjab where the rural areas provided the insurgents with whatever little support that they could manage during the initial period of the insurgency.

The level of support can vary from "active support" to "passive support"; wherein the people can act as early-warning elements, operate radio sets, set off IEDs and lob grenades at unsuspecting security forces in active support, to merely providing food, shelter and moral support in "passive support".

It will also have to be checked if the rot of insurgency has spread to the governmental machinery like the civil service, police and judiciary. This situation is the worst, where no support will be forthcoming in the fight against insurgency.

(b) Psychology of the Insurgent

In order to blunt the cutting edge of insurgency it also becomes useful to understand the psychology of the insurgent. A Company Commander should know the reality and intent behind the facade of impassioned it would only then be possible to predict and counter the movement.

In the present-day context the insurgents are becoming a band of manipulating, fanatical and resolute fighters who are playing insurgency like a well-orchestrated and rehearsed drama. Kashmir, where this aspect is perhaps most easily perceptible, has been the insurgents' psychological game-plan to destroy the credibility of the security forces. This is amply proved by the successful attempts of the insurgents who *"chose to almost never engage Indian troops in an encounter, which was the normal Punjabi tactics, nor did they go in for large-scale massacres of Hindus. They concentrated instead, on firing at patrols and Indian pickets in densely populated, urban areas in order unthinking retaliation. Invariably the people who were killed were innocent citizens ... and every time this happened, militancy would win new supporters to the cause."*[4]

On the contrary it is a vastly different and often underplayed insurgency in Manipur and Nagaland (much to the dislike and frustration of the NSCN-National Socialist Council of Nagaland). The NSCN insurgent is more of a soldier than a diplomat. He would rather not shoot at an officer in civil dress, and when with his wife and children, than break his code of conduct. He is expected to fight like a soldier and a guerrilla. The army which is used to this psychology of an insurgent for donkeys years found the LTTE a more formidablee opponent. As Lt. Gen. S.C. Sardeshpande explains, *"It would be difficult to come by a more motivated, educated, and politicized insurgent militant group than the LTTE. Technical experts, training, wherewithal and innovativeness of the LTTE was of a far higher degree than those of our North Eastern insurgents, whose methods (though effective) were rather primitive in comparison."*[5]

(c) Insurgent Weaknesses vs Security Forces' Strengths

It is imperative to project the weakness of the insurgent and insurgency to make people aware of the pseudonationalists' propaganda and be able to see through the "concern" for them.

Interesting incidents reported by Anil Maheshwari reveal a great deal about insurgent tactics to defame the army.

"Intercepted information on militant communications have revealed interesting details of how these organizations operate. According to army sources, an electronic warfare intercept logged by an electronic warfare detachment on December 15, 1990, revealed that a certain militant had raped the daughter of one Abdul Khaliq. Such incidents too are laid it the door of the security forces so as to show them in bad light and put them on the defensive. Another intercept on January 2, 1991 revealed that militants had planned to wear Army uniforms and get themselves photographed burning a house in village Tikker, Kupwara and thus let the Army defend itself against the charge of arson. Apparently, militant groups have managed to get hold of Army fatigues (condemned) dresses and auctioned Army jeeps so as to enable themselves to pose as regular members of the Indian armed forces. These are typical tactics of psychological warfare."[6]

— The people are most welcome to coerce the insurgents to surrender. The government is willing to help them settle down.

— The army has been prompt in relief programs during periods of epidemics and natural calamities and assistance like medicine is available for the people.

— The government has initiated numerous employment and self-employment schemes including recruitment drives to help settle the people.

- The army is an institution of people from all religions and is an example of unity in diversity.

(d) Futility of Insurgency

It has to be made amply clear to the people that in today's environment it is not possible for any part of the country to disengage itself on the basis of military might. The people must understand they are part of an old civilization where peace and amity have always taken precedence over the rule of the jungle. It must also be stresed that what the table can achieve, guns and terror tacties cannot. The example of the Irish Republican Army is recent and very simile ar to that of the insurgents in India.

- Besides harming the country in general, insurgency harms the people of the state where it is prevalent the most.
- A vast multitude of a generation get killed or maimed for life. The life of a fugitive or a cripple is certainly not worth the futile chase for an inaccessible and unattainable objective and specially so by military means against a better equipped, motivated and organized force.

(e) Publicize Good Work

While it is important to carry on with psychological operations through relentless action and good work, it is equally important to publicize it in order to ensure that it reaches the people and conveys the intehtions of the army and helps them see through the selfish motives of the insurgents. This can be achieved through means that are available at the Company level. These can be:

- Posters and leaflets.
- Lectures and discussions with the citizens' committees and the intelligentsia.
- Occasional talks in schools and colleges.

— Lectures during religious and national festivals.
— Films on actions of the army and the government to help the people in insurgency-affected areas.

(f) Developmental Projects

Rather than mere persuasive talks and posters, the aspect which affects people the most are developmental projects. This is one aspect which besides being good publicity, also has tangible consequences for the people. For a poor man, the availability of rice from a fair price shop despite heavy snow, is of more value than Pakistan's selfish interests in fanning the fires of insurgency. Though most of the projects are capital-intensive in nature, one would find that even at the Company level, liaison with the civil authorities and exploiting local resources can give commendable results. In fact as far as the people are concerned, even the provision of their rights like electricity, water, drainage facilities and food are a favour in the disrupted and stagnant state of civil administration.

The ideal methodology for undertaking developmental projects is to incorporate the people for the manual part of the work. This will increase the direct interaction of the people with the army thereby making understanding of the organization and its aims easier for them. On the other hand the army will also get an opportunity to meet a cross-section of the population and come across sympathetic and cooperative people for the cause of counter-insurgency operations. During the course of these developmental projects, provision of basic amenities like tea and an occasional meal will further strengthen the bond, as sharing of a meal still has high emotional value for the people.

Developmental projects can also be undertaken for religious purposes. Without directly involving oneself with the people in actual work, liaison, manpower from the area, money and moral support can be provided. This will set aside the fears of a genocide on the part of the government hentd the forces as is falsely propagated by security forces as is falsely propagated by the insurgents.

Some of the suggested projects that can be from civil the Company level with some support administration are:

- Rebuilding of schools burnt by the insurgents.
- Completion of construction of masjids and churches.
- Provision of electricity supply. Installation of transformers if the requirement exists.
- Provision of water supply.
- Building of community centres.
- Ensuring smooth functioning of the public distribution system.
- Building of roads and tracks. Planting of trees.
- Provision of low-cost seeds for crops.
- Provision of help for self-employment schemes.

(g) Achieving Goodwill

The very fact that insurgency has cropped in a particular area implies that the vast majority of the people are estranged from the government. Without touching upon the reasons for this, suffice to say that the effect of the hurt sentiments, pride and feelings of the people have to be restored by goodwill. The level of hatred and animosity of some of the people is often an eye-opener for a newly inducted Company Commander. Abdul Majid Dar, a HM militant and adviser-general of the group, has this to say about his impressions about India:

When I went to India I felt I was a member of some enemy country . . . It was always like this when I visited India. If people didn't treat us like enemies, then they treated us like foreigners. I remember during visits to Aligarh and Meerut being asked constantly if I were Irani or

Pakistani. I always felt as if I didn't belong as if I was a second-class citizen.'[7]

It is this misconception that has to be removed. Ideas and fanaticism that has been bred by anti-national discord breeding religious organizations has to be replaced by national pragmatic and sincere efforts of goodwill. Some successful methods that can help achieve this aim are:

— Release of low-level cadres. Low level cadres can be defined as one-time supporters, who may have either been forced or coerced or may have voluntarily helped the insurgents in duties such as lifting of loads, cooking food and carrying messages. Most of these cadres comprise of young boys who live a normal life in the villages as upper ground or over-ground workers. After ascertaining their activities the Company Commander should threaten to hand them over to the police. However this should be resisted and on promises of 'turning a new leaf', the cadres should be released. This should have the sanction of higher HQ.

Notes

1. Boring, Edwin G., *Psychology for the Armed Services*, pp. 485, 486.
2. Cleveld, Martin Van, *On Future War*, p. 20.
3. Giap, Vo Nguyen, *People's War People's Army*, p. 65.
4 Singh. Tavleen, *Kashmir a Tragedy of Errors*, p. 207.
5. Sardeshpande, Lt Gen S C, *Assignment Jaffna*, p. 61.
6. Maheshwari, Anil, *Crescent Over Kashmir*, pp. 171-72.
7. *Ibid*, Serial 4, pp. 221-22.

Human Rights

In a general sense they denote the right of human beings. However, in a more specific sense human rights constitute those very rights which one has precisely because of being a human being.[1]

In the LIC environment human rights gain special significance as the role of the security forces comes under strict scrutiny. It is also because the insurgent's virulent propaganda machinery works overtime to defame the army by concocted cases of human rights abuse to gain sympathy in world opinion. And lastly it gains significance because it serves as an ideal news item to sensationalise the dirty business of war for the media.

"The genesis of human rights can be traced to the emergence of classical idealism. This fiercely defends principles like competitive individualism, private property and market ethics. It upholds and respects the ideals of the individual's liberty, his development and progress through the above principles. The ideas of rule of law, limited government and individualism characterised the seventeenth and eighteenth century anti-nationalist school of political liberalism."

However the main contentions of the twentieth century human rights theory are validity of human rights under legislature, the inherent dignity of humans not being enough. Secondly it receives extended treatment both under Constitutional and International law. Thirdly human rights are vested only in individuals and not in professional groups, social groups, communities, states, nations, etc. And lastly this concept is valid whenever human beings lead a collective life."[2]

The Indian Constitution recognizes the concept of Human Rights in the preamble itself, which begins as, *"We, the people of India..."* Besides this both the Fundamental Rights and the Directive Principles recognise the aspect of human rights.

Whereas these basic principles of human rights are valid for the army, they do not "seem" to matter to the insurgents. MM Khajooria, throwing light on the gruesome acts of the insurgents says:... *"No such compassion, anguish or outrage come to the share of Kumari Girija, who was stripped naked by the Pak-sponsored terrorists and sliced into pieces on the bradsaw. Nor was world conscience aroused and condemnation shouted.*

The other day young Shakeela of Srinagar, who managed to flee to the safety of Nari Niketan at Jammu was lured back to Srinagar, kidnapped, sexually abused and burnt alive."[3]

We must realise, as the upholders of law and order in an LIC environment, that more often than not, the tirade against human rights is one-sided. The security forces become scapegoats on the altar of investigative journalism whereas blatant abuses of human rights by the insurgents go unreported. The excuse, obviously: the fear of death. Senior and respected journalists take the credit for "taking notice of the alleged excesses by security forces in North-East India, Andhra Pradesh, Punjab and Jammu & Kashmir... ."

In this environment despite all the criticism, His Excellency Gen K V Krishna Rao (Governor of Jammu and Kashmir) during his address to the All-India Seminar on Human Rights says: *It has been regularly revealed during the investigation of allegations of so-called custodial killings, that these killings of civilians or militants, almost always take place during clashes between the militants and the security forces. Whenever it is possible, for the security forces to catch the militants, they apprehend the militants. Otherwise there would not have been about 26, 000 militants apprehended in custody and well-looked after, against only about 5, 000 militant casualties so far.*[4]

The deeds of Pakistan in its sponsorship of insurgency and the insurgents themselves in Jammu & Kashmir, by the recoveries of arms and ammunition, killings, and apprehension of insurgents by the security forces vis-a-vis killing of civilians by the insurgents, gives an account of the respect for human rights by these pseudo-human rights propagators.

Destruction by Militancy

	1990	1991	1992	1993	1994	Total
1. Private houses gutted	1145	989	996	997	501	4628
2. Shops gutted	575	174	170	544	133	1596
3. Bridges burnt	45	35	21	47	40	188
4. Schools burnt	138	29	39	71	106	383
5. Govt. buildings burnt	718	161	105	308	157	1449[15]

Weapons Recovered in Kashmir

	1990	1991	1992	1993	1994	Total
1. AK Rifles	1260	2320	3274	2209	1907	10950
2. UMGs	77	48	139	147	112	523
3. RPGs	65	83	97	42	78	365
4. Rev./Pistols	751	853	816	921	826+ 1 Bag	4167+ 1Bag
5. Rocket Launchers	109	27	69	36	49	290
6. Rockets	296	285	250	135	371	1337
7. Magazines	1883	5693	10412	3074	5724	26786
8. Ammunition	187343	249827	263578	324397 +205 Boxes	410406 +80 Boxes	1435551 +285 Boxes
9. Grenades	5136	34134	2274	4663	2283 +4 Bags	48490 +4 Bags
10. Detonators	2634	23501	7843	2985	2201	39164

11. Explosives	232	275	156	2950	35115	396415
	187*	424*	19*	57*	+3*	687*
	40	5	118	4	162	
	slabs	bundles	slabs	devices		Bags"

Toll of Militancy

	1989	1990	1991	1992	1993	1994	Total
1. Militants killed LOC	-	420	374	122	242	155	1313
2. Militants killed	-	130	470	697	1068	1251	3616
elsewhere	19	461	382	634	747	703	2946
3. Civilians killed by militants							
4. Civilians killed on cross firing and maintenance of Law and Order	12	539	524	435	310	220	2040
5. Security forces killed	8	154	168	177	195	160	862
6. Security personnel injured	-	323	520	715	897	661	3116
7. No. of attacks on security forces	-	732	1358	1934	2288	2243	8555
8. No. of	31	475	1126	1765	1628	1577	6602

	cross firing							
9.	Explosions	452	1522	677	757	1005	1012	5425
10.	Incidents of militants firings	92	744	787	859	1026	978	4486
11.	Arsons	12	709	485	678	707	520	3111
12.	Rocket Attacks	-	45	162	105	71	76	459
13.	Abductions	1	169	290	281	349	327	1417*[7]

Toll of Foreign Mercenaries

		Arrested	Killed	Identified
1.	Pakistan/POK	44	156	90
2.	Afghanistan	11	149	52
3.	Bangladesh	164	154	–
4.	Egypt	–	6	–
5.	Sudan	–	8	3
6.	Yemen	–	19	4
7.	Lebanon	1	15	6
8.	Bahrain	2	10	4
9.	Others	–	138	164•[8]

In these circumstances, the role of the Company Commander becomes relevant, as extra restraint and caution has to be maintained during operations in LIC. The following points have proved to be useful:-

— Do not enter a house during a search without the house owner.

— After the search of a house take a written declaration about the well-being of the members and the safety and security of all stores.

— A police person and preferably a magistrate should accompany the army during search operations.

— A dead militant should be photographed along with his captured arms, ammunition and equipment.

— An apprehended suspect should be handed over to the police within 24 hours of the arrest.

— Do not show any laxity in the procedural aspects, specially documentation.

— The details of apprehension, connection with insurgent groups, details of captured weapons, equipment and ammunition should be mentioned in the FIR.

— Get the suspect medically examined before handing him over to the police.

— The FIR is lodged in consonance with the Armed Forces (Special Powers) Act 1958. (See Appendix F).

Notes

1. *Economics and Politics,* Vol. 1, L. 11, No. 1, Mar 1992.
2. Kumar Ashutosh, *Liberalism, Human Rights, Politics and Theory; A Third World Critique.*
3. Address to the All India Seminar on Human Rights.
4. Khajooria, M M, Paper on Human Rights.
5. Sarin, Sat Paul, *Media and Human Rights.*
6. Ibid.
7. Ibid.
8. Ibid.

Part III
Principles of LIC

Principles of LIC

In an environment which produces a variety of situations and gives a range of experiences, it is very difficult to codify principles of LIC as has been done for conventional war. The principles for conventional war have been followed by generals since the advent of warfare: however they were first codified as such, keeping in mind the modern battlefield by Maj Gen John Frederick Charles Fuller in 1921 as part of the field service manual. It was Gen Fuller's firm belief that fighting a war without any firm principles was like *"grammar without an alphabet."*[1] The frustration of the Gen was complete as he wrote in 1916 before the formulation of the principles.

'It is an extraordinary thing (that) whilst every science is based on a few definite principles, war today (1916) should he run on the dice box of Luck... we can predict certain events in war as surely as Darwin could in life, directly he grasped the fundamental principles of evolution. However we had no military Darwin and yet let us hope that the Germans will not discover one.'

The principles of war formulated by General Fuller in 1921 were:

— Objective.
— Mass.
— Offensive.
— Surprise.
— Security.

- Movement.
- Economy of Force.
- Cooperation.

The applicability of these principles can be gauged by seeing the Principles of War of the US Army as given in the current edition of Field Manual 100-1.

- Objective.
- Mass.
- Offensive.
- Surprise.
- Security.
- Mobility.
- Economy of Force.
- Unity of Command.
- Simplicity.

The only difference is simplicity, which has been added to the existing Principles of 1921. As a contrast to conventional warfare LICs are fought against a guerrilla who is difficult to predict and who changes his tactics with changing situations. It is very difficult to codify principles based on which a fluid kind of warfare like LIC can be fought. However, despite the frequently changing environment some similarities and basic guiding principles remain the same. It is these principles which form the basis of the fundamentals to codify the Principles of War upon.

Besides variety that the army has to contend with, it has also to be to borne in mind that the army plays a very vital role in aspects such as public relations, psychological operations, media management and law and order which are not relevant to that great an extent in conventional operations. This broadens the functional repertoire of the army and gives a multidimensional role to it.

Keeping in mind these aspects it is felt that LIC operations should be guided by principles so that they too are not fought like grammar without alphabets. Before I dwell upon the principles, I will like to quote Gen Karl Von Clausewitz who wrote, "The Most Important Principles For the Conduct of War To Complete My Course of Instruction of His Royal Highness The Crown Prince" in 1812. What Gen Clausewitz said about the Principles of War, is also applicable for the principles of LIC today:

The Principles of the art of war are in themselves extremely simple and quite within the reach of sound common sense. Even though they require more special knowledge in tactics than in strategy, this knowledge is of such small scope, that it does not compare with any other subject in extent and variety. Extensive knowledge and deep learning are by no means necessary, nor are extraordinary intellectual faculties. If in addition to experienced judgement, a special mental quality is required it would be after all that has been said, cunning or shrewdness.[3]

The advocated Principles of LIC are:

— Selection and Maintenance of Aim.
— Offensive Action.
— Surprise.
— Mobility and Manoeuvre.
— Flexibility.
— Initiative.
— Security.
— Intelligence.
— Communications.
— Axis of Maintenance.

– Knowledge of Ground.
– Moral Ascendancy.
– Media Management.

The basis of selection of these principles are these two basic questions: *"Is it a battle-winning factor?*, if so, Is it universal for the LIC environment?". In case it is then it qualifies as a Principle of LIC. In the subsequent chapters where these principles are discussed individually, case studies will be taken to illustrate one or more of these principles. Some cases have been suitably changed to illustrate the principles better and keeping security in mind.

Notes

1. Dupuy, T N, *Understanding War,* p. 16.
2. Ibid, p. 17.
3. Clausewitz, Gen Karl Von, *Principles of War,* p. 60.

Selection and Maintenance of Aim

This principle of war forms the basis of a conflict. It not only involves a clear-headed approach while forming the guidelines of conflict in the form of selecting the aims, but subsequently it also involves the more tedious, time-consuming and difficult process of maintaining aims and achieving them. This principle is the embodiment of all the other principles, which if followed keeping this basic principle as the rallying point, will help achieve the objective.

It is essential for everybody from the Company Commander to his junior most sepoy to know and understand the primary aim of their involvement in LICs. Then comes the understanding of the related and perhaps more specific tasks for the area. And whatever the kind, category and degree of LIC environment, the primary or basic aim of the army remains the same. This is *"To create such conditions of normalcy that a democratically elected government is able to function and carry out its governance in a peaceful atmosphere, thereby implementing its social, economic and political aims."* The understanding of this aim involves knowing the related aims and specific tasks. These tasks for an insurgency-prone area can be:-

— Establish control over the area to include area domination.

— Counter-Insurgent Operations.

— Infiltration Control Measures.

— Psychological Operations.

— Civic Action.

— Improve Public Relations.

Peculiar circumstances, the stress of operations, unexpected reverses are at times responsible for creating an atmosphere where the local situation gets the better of the Company Commander and more often young officers and JCOs which makes them forget the primary aims. This results in spontaneous reactions to provocative situations purposely created by the insurgents to bring adverse publicity to the army. Incidents of shoot-outs in public places and deaths of civilians in crossfire are examples. The impact of these stray incidents is extremely detrimental for the achievement of the overall aim. One such incident was the firing by the CRPF at a mob which was moving through the downtown areas of Srinagar with the body of Mirwaiz Maulvi Farooq, who was killed by suspected HM insurgents. Tavleen Singh, a noted journalist, describes the incident is

... On its route lay the Islamia College, which houses the Headquarters of the 69th battalion of the CRPF. Seeing a mob heading towards them, the security force panicked and opened fire, killing 57. The CRPF men claimed they retaliated only after militants in the crowd fired on them.

The next day when they buried the Mirwaiz more than 2,00,000 people collected at the martyrs graveyard near the Idgah. The sound of Kalashnikovs being fired rose above the recitations from the Koran . . .[1]

The damage had been done. A single act of indiscretion had adversely affected the efforts of creating a congenial and peaceful atmosphere. The aim of public relations and psychological operations received a severe setback.

Coming to counter-insurgent operations, the concept of aiming for the objective, maintaining the aim and finally achieving it is the essence of this principle. This particular task will be performed most extensively by a company in LIC. For the completion of this task, the coordination of the Company Commander, his command and the task form the three constants of which the Company Commander should so influence the troops that the task is performed.

SELECTION AND MAINTENANCE OF AIM

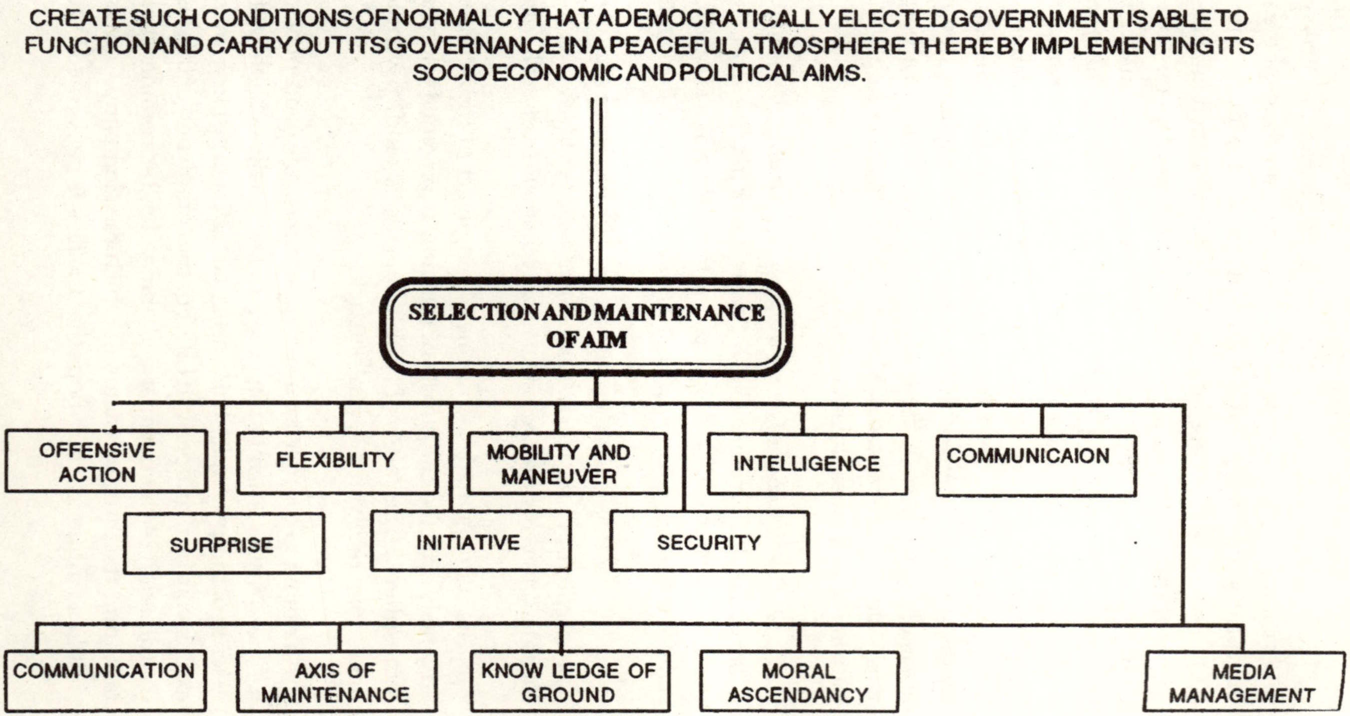

Figure 3

PRIMARY AIM AND SECONDARY TASKS

CREATE SUCH CONDITIONS OF NORMALCY THAT A DEMOCRATICALLY ELECTED GOVERNMENT IS ABLE TO FUNCTION AND CARRY OUT ITS GOVERNANCE IN A PEACEFUL ATMOSPHERE THERE BY IMPLEMENTING ITS SOCIO ECONOMIC AND POLITICAL AIMS

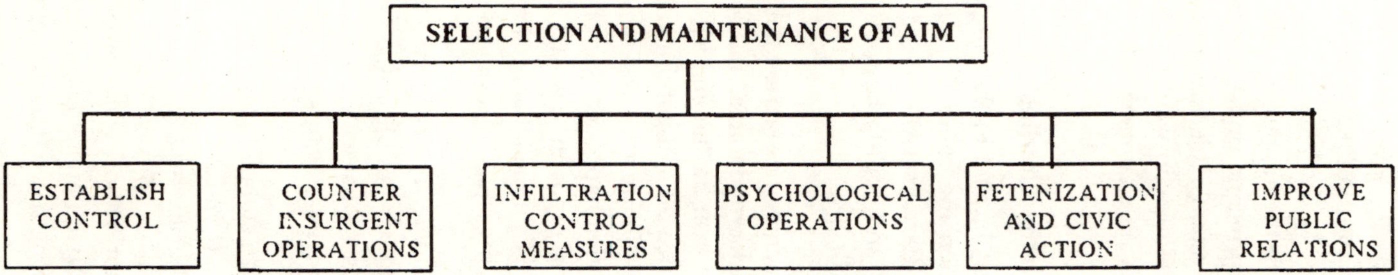

Figure 4

A case study from the jungles of Vietnam will illustrate the employment of the principle under discussion. It is a specific example of a ci operation. This first person account has been narrated by Stanley Goff who went to Vietnam as an ordinary soldier and was a machine gunner and a fire team leader at the time of this incident.

Pitched Battles with the 'Charlies'

The company was located in the jungles of Vietnam. Based on specific information the company was moving along with Armored Personnel Carriers (APCs) and tanks while moving across a paddy field and approaching the edge of a jungle, an APC was hit by a shell.

That day as I remember we started going real slow. We were riding along in a thick woody area, and all of a sudden, out of the clear blue sky, we heard a "boom bam ... dididididi". The APC stopped and we jumped off and got down besides the tanks. We were looking, trying to figure what was happening. It was way in the back ... coming. What happened was of course they knew we were at one of the carriers got hit. It wasn't a bad hit, but five guys on top of it got blasted by a shell that hit the top of a carrier, so we had to stop and wait, my squad just stayed put.

Slowly we found out what happened. One guy got his arm almost blown off; another got hit in the eye. Damn. So we said "Well this the hit. This is what we've been riding for so goddamn long for." We were moving towards a major conflict.

I guess the caravan stopped about an hour. Then we started moving again. We knew we had to get the assault in the backs of our minds, because the next tank to get hit could be us. . .

. . . I never will forget how we approached it, the tanks and APCs quietly lining up in parallel formation. The rice paddy was about two times the length of a football field and about a football field in width ...Within fifteen minutes interval we stopped and lined up at the rice paddy. Then the word came, "All right, dismount and stay at the back of the carriers." So the men started to climb down the sides of the vehicles. All of a sudden the carriers started reconning by fire. They just started firing at the wood line, "boom, boom", with these big tank guns, just tearing the wood line all up. Man the whole damn wood line opened up, "Boom didididi Wham Wham ..." Rockets. I heard guys getting hit from over to my left. I heard a tank get hit. I didn't know how bad.

Now my mind was jumping. By this time everybody was reconning by fire. I was firing back automatically even while this barrage was coming in. Everyone was standing up there doing nothing but firing like hell. Pretty soon we were told, "Back up, back up, back up." I thought to myself, "God damn. Shit, ... it's hell over there ..." This regiment had probably left a suicide battalion over there to knock the shit out of us, so that the rest of the enemy could go on and do what they had to do. We pulled back into the opposite wood line.

When I pulled myself together, I was looking around for my men. They were really shaken up. I could see the shock in their faces — no blood at all in their faces. They said, "Hey man, are we gonna go across to that wood line?" "I said, "Yup, I think we are." "Wow man that suicide." "Could possible be man." I didn't know what was going on towards the other end of the column ... After I pulled back into the wood line I made sure my weapon was clean. That was what my squad saw me doing.

Soon we heard a helicopter come in. They were medivacking a guy. One of the touts was blown away; took a direct hit ...and then we found out that we were actually going to assault that wood line "Assault on the wood line!", A lot of guys were saying. I wasn't saying anything ...

I got my weapon all cleaned up and made sure that all my men were around me, and I didn't do much talking. I said, "OK men, primarily what I want you to do is just stick by me, OK? When I call for that ammo, I want you to have your ass right there — you got it?" "I got it, Goff, OK". "OK fine just keep your head down man." "OK ..."

You see a gunner needed an ammo bearer that was not so worried about his own head, that he couldn't effectively feed the gunner the ammo. I would be blowing lead out of that pig so quick I'd go through a belt in ten seconds, needs a man to be able to hand me the ammo. He didn't have to stick it in the weapon. I did that. He just simply handed it to me, and I hopped it in there. I could do it faster than he could anyway.

As we got ready to go back up to the wood line, the NVA stopped firing, waiting for us to charge. It was very quiet over there. Then the tanks moved out and started firing as they went, the NVA returning their fire. We all started moving out too, walking at first just behind the tanks, letting them do all the firing. As the fire came in, I heard it hit on top of the tank that I was behind — dding, dang, ding. As the tanks started going faster and faster, they cut us loose as they got ahead of us. Obviously as that cover pulled out about ten feet ahead, we started lowering ourselves and we started firing. As they finally pulled away from us, we all hit the dirt, out in the middle of the rice paddy and started inching our way towards the dike. Then we were all running towards the first dike with the tanks forty feet ahead. We couldn't fire too much because they were still ahead of us. So we mostly kept our heads down and moved towards the first dike, about two feet high-high enough for protection. As infantry our job was to take care of the NVA who might have moved on foot to attack on tanks and the personnel carriers from the rear ...

I was at the dike firing like hell with Emory right with me just handing me the lead. He said, "Hey, Goff, I'm out of lead. What do you want to do?" "Don't worry I got enough right down here", and I was still firing. "What I want you to do is go and get all the ammo from the other guys down at the other end of the company. Find anybody that's got ammo, just get it,..."

Emory and I were running down this rice paddy field firing. The guys would tell me "Hey Goff, right here, right in there, man". I would sit down between the two guys and blow out where they thought they were getting heavy concentration of fire. Then Emory and I would run into another area along the dike. When Sgt Needham hollered, "Goff, Goff, over her, man, I got thirty or forty of them right there, right there", I'd fire right where he told me to fire. Those were the thirty or forty enemy I am accredited with in that area. Emory was not with me. I told him to stay while I ran over and was firing my ass off in this particular area, so he started firing his M-16 too.

By now we were in the middle of the paddy at the first dike, which we went over. We cut down the body of men so well, knocked down their fire power, that we could move on towards the second dike at the end of the paddy firing steadily. After we got to the second dike, I went on firing for about fifteen more minutes, but then my pig fell apart. It just blew up in the air, like it did earlier in the area. This time the barrel did fine, but the pins came out of the side of the weapon. It just got too hot, and when it expanded, the pins and the locks and the keys that held it in place were no longer workable, and the pig just came apart ... I was slumped on my knees at the second dike just staring. The second dike was almost at the wood line. With us being at the wood line and me sitting there exhausted and with the area completely quietened a few of the other squads started running into the wood line, crouched, searching, looking, weapons at the ready ...

I was groggy, but we had to move out; so what if I was groggy. I could hardly get on top of the Goddamn tank. I was so weak. Sitting up there I saw all these bodies, or parts of bodies — hands, arms so much so that it was making me sick to see all these bodies lying on the ground. I realise that it could have been me there ...

We dismounted and 1 walked about ten or fifteen feet up to the porch and collapsed. "I can't move." It was no laughing matter then. I was conked out on the ground. And I stayed there. My sense at that time was that I had just been in a helluva battle and that I had done nothing more than anybody else did, that I had done nothing outstanding; but that I was alive, I had survived.[2]

The vivid and thrilling account proves beyond doubt the Principle of Selection and Maintenance of Aim, wherein the Company despite the heavy opposition, being fired at whilst doing firing, and suffering casualties, came right back and wiped out the insurgents.

The incident is also a graphic account of a true-life scenario, the feelings of men and the humble, honest and down-to-earth leadership of a squad leader.

Notes

1. Singh, Tavleen, *Kashmir a Tragedy of Errors,* pp. 158-9.
2. M Malone, Col Dandridge, *Small Unit Leadership.*

Offensive Action

"...The great distinction of this game is that it truly, when well played determines who is the best man. Who is the highest bred, the most self denying, the most fearless, the coolest of nerve, the swiftest of eye and hand. You cannot test these possibilities wholly, unless there is a clear possibility of the struggles ending in death."

— John Ruskin in the *Chorus of Wild Olives.*

"Who Dares Wins", the SAS motto aptly summarises the spirit behind this principle. Contrary to general belief, the application of this principle is not restricted to counter-insurgent operations alone. It encompasses in its preview the entire gamut of LIC operations and includes aspects as diverse as psychological operations, media management and public relations. Even before counter-insurgent operations are launched, an ideal offensive policy will root out the insurgent movement without the use of force. This political initiative is a more powerful and effective weapon of Offensive Action than the mere use of arms.

This thought process is not an evolution of the recent past but has its origins in the history of diplomacy, statesmanship and warfare. Sun Tzu, a Chinese ruler who lived more than 2,400 years ago — at roughly the time of Confucius, had the rules of war worked out even before the time of Alexander the Great. Sun Tzu sums up offensive strategy briefly yet clearly as:

Thus what is of supreme importance in war is to attack the enemy's strategy. Next best is to disrupt his alliances by diplomacy. The next best is to attack his army. And the worst policy is to attack cities ...

Those skilled in war subdue the enemy's army without battle. They capture the enemy's cities without assaulting them and overthrow his state without protracted operations. Their aim is to take all under heaven intact by strategic considerations. Thus then troops are not worn out and their gains will be complete. This is the art of offensive strategy.[1]

When compared with the modem day LIC, this implies that the ideal method of combating a potential LIC, is to defeat the enemy strategy of subverting the minds of the people, which is recognized as the first stage of LIC, subsequently isolating the insurgents by winning over the people and cutting aid from the supporting country in the second stage. In the event of failure of this strategy, diplomatic offensives are started, to make support to the cause of the LIC and specifically insurgency untennable. The third stage is of military and social operations specifically against the insurgents. The last stage results in a civil war where military action is ordered on a mass uprising against the government.

Having had an insight into offensive strategy or action at the macro level, let us now consider specifically the third stage of this strategy that is counter-insurgent operations and psychological operations.

The effective domination of the area is the crux of Offensive Action at the company level. This involves extensive planned movement in the area of responsibility. Often this exercise is considered an avoidable torture and is termed as "Jungle Bashing" but what the detractors of this operation fail to understand is that an insurgent remains as far from the post and area of responsibility as he is forced by the actions of the army. Given a chance he will like to prove his superiority militarily as well as psychologically to the people by jitter actions, ambushes and improvised explosive devices. The area domination operations not only deter his action but produce a sense of security and normalcy in the minds of the people. This also gives confidence to the people to visit the post and give usefull information.

Offensive Action throws the plans and composure to plan of the insurgent, in a quandary. The constant movements as reactions to the actions of the army forces the insurgents to follow the tactics of "reaction" rather than "action". The pressure exerted due to these offensive actions is one of the major reasons for the insurgents to surrender. This was evident from the large-scale surrenders of the ULFA cadres in Assam who could no longer live the life of fugitives because of constant harassment by the army.

The point to be borne in mind is that operations should be launched on specific information rather than as an offshoot of frustration. This is the aim behind calling operations people friendly. Though this may seem contrary to the operations of area domination, it certainly is not, as area domination does not involve random searches at odd hours of the night without any specific information.

Counter-insurgent operations often result in the insurgents getting away once contact is made. This is where the offensive spirit of a commander and his command is displayed. At no cost should contact be broken with retreating or fleeing insurgents after it has been established. Whereas the escape of the insurgent is a tremendous morale-boosting factor for them, it hurts the pride and self-confidence of the company. Conversely an insurgent caught after a chase or a counter-action gives a perceptible psychological edge to the army.

Offensive spirit is easier demonstrated when the insurgent is on the run. However, the best demonstration of offence comes when the chips are down, when men are under fire and when all seems lost. It is at this stage that years of training and strengthening of the bond between the Company Commander and the Company comes into play. This aspect of Offensive Action thus forms the crux of this Principle of LIC. An incident which occurred during the Second World War between a French patrol led by Lt Woodrow W Millsaps and a German party is very similar to the numerous incidents of LIC which either result in the security forces preferring to follow the dictum of "discretion is the better part of valour" or "uninhibited offensive action".

In June 1944 Lt Woodrow W Millsaps led a patrol from Hill 30 down the causeway leading to Chet du Pont. It was a night action during operations along the Merderet River in Normandy. Though the men had volunteered for the duty, all were in a particularly desperate state of mind before the start. For three days the battalion had been surrounded by enemy forces; within the perimeter men were dying for want of plasma and the wounded were suffering more acutely because of the lack of water; their cries of distress had so demoralized the able-bodied men that Millsaps welcomed the chance to get away from the hill.

At the foot of the hill an enemy machine gun opened fire on the patrol but the bullets went high. The men broke and 'ran like dogs'. Millsaps and a Sgt beat them back with physical violence. After they were again collected, Millsaps lost almost an hour, alternately bullying and pleading with them before they would go forward.

At last they charged the enemy, closing within hand grappling distance. The slaughter began with grenade, bayonet and bullet. Some of the patrol were killed and some wounded. But all now acted as if oblivious to danger. The slaughter once started could not be stopped. Millsaps tried to regain control but his men paid no head. Having slashed every German in sight they ran on into the barns of the French farmhouses where they killed the hogs, cows and sheep. The orgy ended when the last beast was dead.

"Millsaps then called for volunteers to outpost the position and accompany him across the river. All but one man held back. Millsaps started in with this lone willing soldier. The man soon began to fall behind. Millsaps asked. "Something wrong with you?" He answered, "I don't think so." Then Millsaps stripped the mans jump jacket away and

found six bullet holes in his upper right arm and shoulder, the soldier had not been aware of his wounds until that moment. The soldier collapsed. Millsaps continued on alone."[2]

OPERATION KONTAKKARANKULAM was launched by the Army against the LTTE in the Vavuniya sector. This operation was a company level operation and proved the offensive spirit of the Company Commander, Havildar Bhim Bahadur Thapa and the company. The operation also gives an example of flexibility and manoeuvre despite heavy enemy fire.

Operation Kontakkarankulam

General

Battalion 'A' was inducted in Mannar Sector of Sri Lanka in January 1988. The battalion moved to Sri Lanka directly from an area of insurgency where it had been operating since 1985. On 23 May 1989, it was redeployed in Vavuniya Sector, with a company each at Omantai, Puliyankulam and Putur, while Battalion HQ with one company was located in Puliyankulam. 'C' Company was deployed at Omantai alongwith a company of Battalion 'B'.

Background

Omantai was a staging camp for the East-West move of LTTE militants. The company of Battalion 'B' was generally employed in Road Opening and Convoy Protection duties and as such dedicated itself for gaining intelligence/carrying out tactical operations. Exploiting this situation the LTTE extended its influence in the area significantly. 'C' Company, Battalion 'A' was therefore tasked to reinforce the Company of Battalion 'B', dominate the area and hunt the Tigers in that area.

Reconnaissance and Intelligence

Immediately on its arrival at Omantai, 'C' Company carried out quick and extensive area familiarization by sending joint patrols and ambushes. EPRLF (Eelam People's Revolutionary Liberation

Front) guides were also incorporated in these missions as they had intimate knowledge of the terrain and modus operandi of the LTTE militants. Within a short span, adequate information was gathered to carry out operations in that area.

Information Gathered by the Company

The LTTE militants were operating from general areas Kontakkarankulam, Pekkirukulam, Vimankullu and Nallikulam. Their known modus operandi was to lay ambushes to harass/destroy IPKF patrols and convoys. For food, shelter and taxes they depended on the areas mentioned above and the civil transport operating on Vavuniya-Jaffna Highway. LTTE militants never maintained static camps in IPKF areas of responsibility. Most of their main camps were in the Sinhala/Tamil border areas with the sea coast on the east to facilitate quick getaway. Militants were generally in the area north of Nallikulam but their exact locations were not known. Militants were generally operating in groups of 10 to 12, with 5 to 6 weapons.

Proposed Modus Operandi of Own Troops

It was decided that ambushes by own troops should be laid on the routes on which their move pattern was observed. They should be in position preferably an hour before first light. After withdrawal/springing of the ambushes, selected search and destroy missions to be carried out. Use of IEDs should supplement own ambushes to inflict maximum casualties. Strength of own patrols/ambushes should be commensurate with the militants' strength, and distance from own posts.

Operational Plan

Task

'C' Company less a platoon was tasked to carry out search and destroy mission, in general area North of Ilamaluthankulam and South of Nalkulam and lay multiple ambushes on the most likely approaches of militants from June 21, 1989 to June 23, 1989.

Troops to Task

Two platoons ex 'C' Company, Battalion 'A' alongwith two EPRLF cadres were to constitute the main ambush. One platoon ex 'C' Company was kept as a QRT. Two 81 mm mortars ex Battalion 'B' located at Omantai were to provide fire on call. Ranjit and Akbar helicopters were available on call.

Outline Plan

Two platoons ex 'C' Company, Battalion 'A' were tasked to lay ambushes at Road-Track Junction and Road Bend respectively. Ambushes were to be effective by first light on June 21, 1989. Ambushes were to remain in location till 1100 hours on June 21, and thereafter carry out search. On June 22, ambushes were to be laid on Track Nallikulam-Road Bend and were in situ till 1200 hours. Thereafter search of the remaining area was planned cross-country. The operation was planned in conjunction with the road opening being carried out on 21 June 1989, by a company of Battalion 'B' to achieve security of our own plans. Troops were to be self-contained for 48 hours.

Execution

'C' Company Battalion 'A' less a platoon left their base at Omantai at approximately 0200 hours on June 21 for the assigned ambush sites in the areas Road-Track Junction and Road Bend. After moving cross-country through thick vegetation and jungles, the columns reached the general area of Road-Track Junction at approximately 0415 hours from where number 9 Platoon was to leave for their assigned ambush site. As number 8 Platoon was in the process of occupying the ambush site, the outbreak of dawn was shattered by sudden and intense automatic fire which came from a distance of 20 to 25 metres. Due to poor visibility, it was difficult to locate the exact area of the militants' ambush. Own troops immediately returned the fire in the direction from where the militants had opened up. The firing continued for about ten minutes. In the meantime, the visibility conditions improved a bit. The Commander saw three militants lying in a pool of blood approximately 25 metres away with their weapons. Another 8 to 9 LTTE militants were firing from behind the tall and thick trees. Appreciating the futility of continuing operations from the same position he ordered number 8 Platoon to provide a fire base and engage the militants. He, along with number 9 Platoon led the charge at the militants from the North-Western direction. On being surprised, the militants ran helter skelter. Own troops taking advantage of this, started firing from hip positions. One round of 84 mm RL was also fired on the fleeing militants. During this daring assault on the militants, Havildar Bhim Bahadur was hit by an automatic burst on his chest. Immediate evacuation was asked for but Havildar Bhim Bahadur breathed his last before the helicopter could land.

Number 8 Platoon was ordered to pursue the fleeing militants. This column moved South of Naillikulam and halted. Further move was not considered desirable due to suspected location of a major LTTE camp in that area.

Number 9 Platoon which was by now in the area of Kontakkarankulam drew fire from another group of militants from approximately 400 metres North-East of Kontakkarankulam. During this the Company Commander and two Other Ranks were injured but they continued operations after having got first aid. Number 9 Platoon returned the fire immediately. Heavy and accurate 81 mm mortar fire was brought from Omantai on the militants' position. At this juncture a sortie of Ranjit helicopters was asked for. It came and carried out strafing on the fleeing militants. Number 8 Platoon which was in the area of Nallikulam was ordered to establish stops in that area, intercept the fleeing militants and destroy them.

Thorough search of the area was thereafter carried out. Besides some blood trails nothing else was found. The troops returned to Omantai by late afternoon on June 21, 1989. Successful operations were launched subsequently based on valuable information deciphered from the captured documents giving details of their camp.

Achievements and Losses

In this gallant action Havildar Bhim Bahadur Thapa died due to gunshot wounds. He was decorated with a Vir Chakra (posthumously) for displaying valour of a high order.

The militants suffered 5 dead on the spot and 8–9 injured militants succumbed to their injuries according to militants' radio intercepts. Villagers had seen 15 bodies being cremated at Naval on the same day. The following arms, ammunition and documents were recovered:

Rifle AK-47 - 1

Rifle G-3 - 1

Rifle M-16 - 1

Ammunition (all types) (rounds) - 407

Mines (all types) - 4

Radio Set - 1

Compasses - 3

Maps and sketches - 4

Photo albums - 6

LTTE Casualties

Dead - 5

Injured - 8 - 9

Losses Own Troops

Dead - 1

Based on the information extracted from the documents, a number of operations to apprehend and destroy the militants were launched which met with a fair amount of success. Intense shelling of the camp areas of militants was carried out in which militants suffered heavy casualties which was confirmed by the radio intercepts.

Notes

1. Sun Tzu, *The Art of War*, p. 105.
2. Marshall, S.L.A., *Patrol*.
3. *Prativirodhi*, Apr 1992.

Surprise

Surprise results from going against an enemy at a time and/or place or in a manner for which he is unprepared. It is not essential that the enemy be taken unaware but only that he became aware too late to react effectively.

— US Army FM 100-5 (1980)

Traditionally surprise has been the weapon extensively used by the insurgent to defeat, demoralise and disgrace his more fancied opponent. Achieved through a vast variety of operations, surprise as a principle remains a nightmare for security forces the world over. It is this particular principle which if wrested from the insurgent takes away the advantage of guerrilla warfare from him and exposes him as a conventional rival or a hiding fugitive.

Surprise as a principle of LIC does not only deal with the operational level. In case this mistake of taking surprise merely as an encounter-winning factor is made, the insurgent may be taxed but the brain behind the movement continues to work. Surprise as a principle should therefore aim for the mind which if numbed by the continued use of this principle will cripple the will to fight. But surprise should form the crux of each policy announced, each announcement made, each decision taken, each discussion held and, last but not the least, each operation planned.

It should form the basis of psychological operations, media management and public relations. This multi-pronged utilization of surprise will throw an unexpected situation time and again to the insurgents, thus keeping them guessing.

The principle of surprise should be coupled with other principles like Intelligence, Deception, Security and Manoeuvre to remain one step ahead of the insurgents while planning an operation. Sun Tzu has said, "That what allows wise men to get that, which lies beyond the reach of common mortals is precognitive (what might happen)."[1] Deception leads the insurgent to believe what is not. It creates circumstances by which the insurgent does what the army wants him to do. Security ensures that plans of the army are not compromised and secrecy can be maintained for the duration of the operation. Lastly manoeuvre ensures that intelligent and rehearsed switching of resources creates the element of surprise. All or some of these principles have been recognised as essential ingredients for producing surprise. For Clausewitz the matter is quite simple. *"The two factors which produce surprise are speed and secrecy."*[2] Other thinkers and nations go further to prove the point. The (erstwhile) Soviet operational writing talks in terms of *"swiftness of troop operations, use of night and unexpected use of nuclear weapons"* amongst others. The last is not really an element that will be useful for LIC, and in the changing world order, even conventional war. The Americans stress similar aspects. *"Factors contributing to surprise include speed and alacrity, deception operations of all kinds, variations of tactics and method of operations and operations security."*

Having discussed the principle of surprise with a view to gaining a broader perspective let us see how surprise can be achieved at the company level. It will be seen that the suggested measures are all offshoots of the four basic principles of achieving surprise, i.e. *'Intelligence, Deception, Security & Manoeuvre.'*

Night movement is still an effective means of achieving surprise and will continue to remain so until the insurgents achieve the capability of an effective night vision capability which at present is positively absent. The aim of surfacing where least expected is bound to achieve surprise. However, the basics of field craft and battle craft have to be perfected to achieve this. It is also essential to have good knowledge of ground. The use of a reliable guide is an added asset who will almost certainly know the area better than the security forces.

Deceptive routes can be taken to reach the target. The route taken can be at a tangent to the target area so that the locals are not alerted about the proposed operation. This will take more time to reach the proposed area of operations, therefore allowance should be given for the extra distance to be covered.

Similar to this, two or more parties can be sent to confuse the insurgents and their supporters about the actual area of operations. These can either be launched from the same location or from separate posts. However, care should be taken to avoid clashes and mix-ups amongst our own troops.

Another means of achieving surprise is to purposely spread false information. This can be done by a variety of means. False transmission is an effective way of achieving desired results if it is known that the insurgents are listening on the net. However, this requires careful and systematic planning. Not only should the Battalion HQ be in the picture of the deception plan, but the neighbouring companies and paramilitary forces should also be in the picture if they are on the same net. Others means are spreading information through own sources as 'leaks' or 'loose talk' by an irresponsible soldier.

Use of small parties for operations based on specific information are likely to achieve surprise. A large number is quite likely to produce lethargy and a feeling of remaining detached from the forward section, which results in giving away of surprise. The risk involved in using small parties is more than compensated than by the chances of success.

There may be circumstances where operations in civil dress

are allowed. In case they are, then these can very effectively be used to achieve surprise. It must obviously be borne in mind that these should be conducted with the help of civil vehicles and not in our own "civil painted vehicles". These operations require expertise and can prove counter productive in case precautions are not taken. The operations must preferably be carried out with suitable equipment to include weapons and radio sets.

A good intelligence network provides the basis of surprise. Timely, authentic and actionable intelligence can help to act, to achieve surprise against the unsuspecting insurgents rather than *reacting* to information.

Monitoring of transmission of the insurgents has proved to be very useful in gaining valuable information. Captured sets and at times specially provided radio sets can be used for this purpose. A source who knows the local language can be used for despatching messages. The ideal method of extracting accurate information is to first record the transmissions and then get them translated. This reduces the unnecessary reliance on the memory of the source to translate the message. It also helps to decipher codes that may have been used during the conversation.

Speed is another factor which can achieve surprise. When a party moves faster than it is expected to, and reaches its target areas, it is bound to achieve surprise. However caution must not be lost sight of if the circumstances indicate the presence of insurgents.

Ultimately the ability to wait, to be patient helps achieve surprise. Remaining hidden in a hide and waiting for an unsuspecting insurgent can be a very taxing and frustrating experience. Whoever is able to overcome this will certainly achieve surprise.

An operation conducted in OP PAWAN clearly illustrates the use of the principle of surprise. Alongwith this the ambush party commander's maturity and cool-headed approach based on sound intelligence about the insurgent tactics bore fruit.

OP Merry Lane

Background

In May 1988, sources were tasked to identify tracks along which the Tigers were moving. Intelligence sources by then had already confirmed that the Tigers, particularly their leadership, was moving on vehicles on specific tracks. The identifying of tracks that were being used was a long and laborious process which over the years paid handsome dividends.

Intelligence

'Capricorn' sources had already indicated that there was movement along an unmetalled road that ran from Puliyankulam to Omantai. Therefore the ambush party commander planned to lay an ambush with his party on this axis.

Terrain

The general area, like elsewhere, in Wanni was densely forested. However, the area around the ambush site was innocuous. It was an area interspersed with scrubs, jungle and an open area, thus affording adequate cover, early warning of the militants' approach and a good field of fire.

Plan

Puliyankulam is approximately 11 kms from Omantai. The party commander realised that any build-up on the Omantai post by the commandos would be definitely reported by the sources of the Tigers. Therefore he decided to rely on deceit to reach the ambush site. He decided to employ a civil truck which belonged to a 'Capricorn' source which regularly ran between Kankarayan. He planned to move cross-country during the hours of darkness and lay the ambush before first light.

Approach to Ambush Site

The ambush party moved from Puliyankulam on Oct 9, at 1900 hours in the civil truck. The troop quickly dismounted at Vilkvetikulam and the truck unsuspiciously continued its journey. Moving with stealth through the extremely dense forest the party established a commando base around 0001 hours on the same night. The troop halted for a short while at the base, and once the moon was out at about 0300 hours (on June 10) they abandoned their base.

The troop painstakingly moved slowly towards the ambush site. During the entire mission they used only PT (light, canvas) shoes so that the telltale imprints of jungle boots would not give them away. In fact while walking on the tracks the troops moved bare foot like the locals of Wanni.

Siting of Ambush

The details of the siting are given in the sketch attached at Figure 5.

Conduct of Ambush

At approximately 1030 hours on Oct, 10 scouts informed the commander of the approach of two motorcycle-borne militants. The party commander in his wisdom let the militants go past appreciating that they were the vanguard of a larger force.

At around 1035 hours the scouts reported the approach of four militants on two cycles. The militants were clad in shirts and 'lungies'. The scout also identified one of them as Mohan, a group leader. The scout also simultaneously said that he could hear the sound of a tractor. The ambush party commander was now certain that definitely an important leader was moving towards his ambush site.

At 1040 hours the scout confirmed a tractor with a trailer crossing them, carrying armed militants.

LAYOUT OF AMBUSH

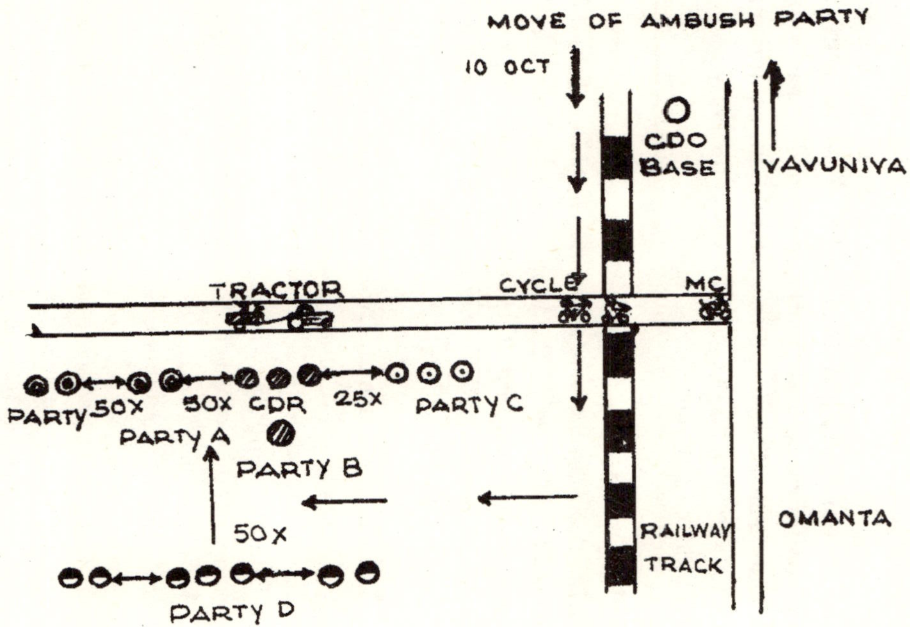

When the tractor reached the centre of the site, the firing party sprang the ambush. The tractor driver was hit in the first volley, the tractor thus went out of control and lurched towards the ambush site. To the credit of the Tigers they immediately assaulted our ambush position, but the ambush party demonstrating professionalism of an outstanding order held their ground and cut down the militants with precision fire. They accounted for all the twelve militants who were on board the tractor trailer. This was the most outstanding ambush (till then) of the IPKF. Exercising total control the ambush party conducted a search of the dead militants. All their weapons, ammunition, equipment and one Icon radio set was recovered. Thereafter the party withdrew to an IPKF post at Omantai.

Results

The Tigers lost twelve of their cadres including the Nedukani Sector Commander, Maj Dadhi Bala, Mannar Sector Commander Simon, and Parthasarathy, bodyguard of Mahataya Shri. Our troops sprung this spectacular ambush without any casualties. The troop also expended little ammunition to achieve this success.[3]

The effect and use of surprise as a principle of LIC has also been effectively used more than a century before OP MERRY LANE was launched. It is ironic that the scene of this action was also Chechnya and the advancing force was Russian led by the Russian General Golosojef. This incident of 1840 had also made the Russians pay a heavy price before they could subdue the numerically inferior adversary.

In 1840 the Russian General Golosojef, advancing with a force of about 5, 500 men against the guerrillas of Chechnya, fell into a cleverly prepared trap and suffered heavy losses, although ultimately successful. The enemy had arranged a most skilful ambush along the banks of the little river Valerik and in the neighbouring thickets. The Russian troops were on the march, suspecting no evil, and without elaborate precautions. Suddenly a heavy fire was opened upon them when they were quite unprepared, and by its unexpectedness nearly caused a panic. In the end after a severe hand-to-hand encounter, the regular troops proved victorious, but no before they had lost large numbers of killed and wounded at the hands of a numerically insignificant adversary.

Notes

1. Sun Tzu, *The Art of War.*
2. Clausewitz, *The Art of War.*
3. *Prativirodhi,* March 1994.
4. Callwell, Col C E, *Small Wars.*

Mobility and Manoeuvre

Move when it is advantageous and creates changes in the situation by dispersal and concentration of forces. When campaigning, be as unfathomable as things behind the clouds; when moving, fall like a thunderbolt. When you plunder the countryside divide your forces. When you conquer territory, defend strategic points. Weigh the situations before you move. He who knows the artifice of diversion will be victorious. Such is the art of manoeuvring.

— Sun Tzu

Manoeuvring involves the strategic or tactical positioning of troops to a place of advantage. It requires mental flexibility, quick thinking and initiative to be able to locate the troops and surprise the insurgents. Similar to the principles of surprise and flexibility the basic art of manoeuvring too is the hallmark of insurgent tactics. However to challenge him and beat him at his own game it is essential to out manoeuvre him and gain a position of advantage to hit him from.

The insurgents will have to be out manoeuvred on the psychological and the military fronts to ensure a complete rout. His vicious propaganda to defame the army and to project it in a poor light will have to be answered, and more importantly, initiated, to keep him on the defensive. Militarily he will have to be out manoeuvred and beaten at his own game of hide and seek.

It is a recognized fact that the insurgents are able to move faster than the security forces in their area of operations. This is due to their light loads, lesser number, physical conditioning suited for the area and knowledge of the terrain. These very factors will have to be worked upon to ensure that we are able to match him at his own game. This is easier said than done. It not only requires hard work, but it also requires sustained efforts to improve the standards of the company. In urban areas, mobility of the sub-unit will gain added significance to achieve the conditions for swift manoeuvrability.

The other principles of LIC which directly affect this principle in achieving its potential are Communications, Initiative, *Flexibility, Intelligence and Vision.*

A sound and effective communication network helps accurate passage of orders and feedback. This ensures the speedy execution of orders which is very essential for the principle to be useful.

Manoeuvrability requires initiative to take the chance and risk of stretching forces and thereby outwitting the insurgents.

Flexibility is the basis of manoeuvrability. In fact there is an overlap in the basic fibre of these two principles, in the sense that neither can be used without the other automatically being used. Yet they are different in their basic employment.

Intelligence gives the data to work the plans on. Without accurate intelligence, a Company Commander may end up moving to a position of disadvantage and lose initiative and surprise.

The vision of a Commander gives him the ability of manoeuvre to a place of significance in the future. It is essential to be able to look beyond the nose and foresee the turn of events.

A Company Commander will require a QRT suitably placed to be able to move to a position to achieve his aim. The location of this reserve should be selected with due thought and deliberation.

Physically fit troops can reach the assigned place at the given time. This requires emphasis and forms a basic tenet for the principle.

An operation conducted in the Rangwar area in the LIC environment brings out the principle of manoeuvrability and concentration.

OP Dhanush

On the night of Dec 07-08 92 an ambush party, comprising two officers and sixteen other ranks ambushed a large group of 50 heavily-armed, Pakistan-trained terrorists who had infiltrated into Indian territory. Rangwar is located on the Eastern slopes of the Shamshabari Ranges because the area is highly avalanche-prone. In winters, the posts are vacated for some time during December/ beginning of January depending on the level of snow, and re-occupied in March/April. During this time, no major infiltration is possible as the passes are blocked by heavy snow which makes movement with loads difficult.

With the onset of spring and the thawing of snow, it becomes a race between the Army and the ANEs (Anti National Elements) as to 'who gets there first'. The terrain configuration enabled the ANEs to plan their infiltration with relative ease. The method generally adopted by them was well-thought out with the help of guides.

From our own side of the LIC, guides would cross and contact the nearest Pak post. They would then be assigned groups of infiltrators by the ISI officers who control the militant organisations. On the designated day, the infiltrators would be staged forward to one of the forward Pak posts. The group would wait at the hide, while the guides would cautiously work their way up, reaching the crest sometime between 1600 to 1700 hours. The location of our ambush parties and patrols would be

OP DHANUSH 7/8/ DEC 92

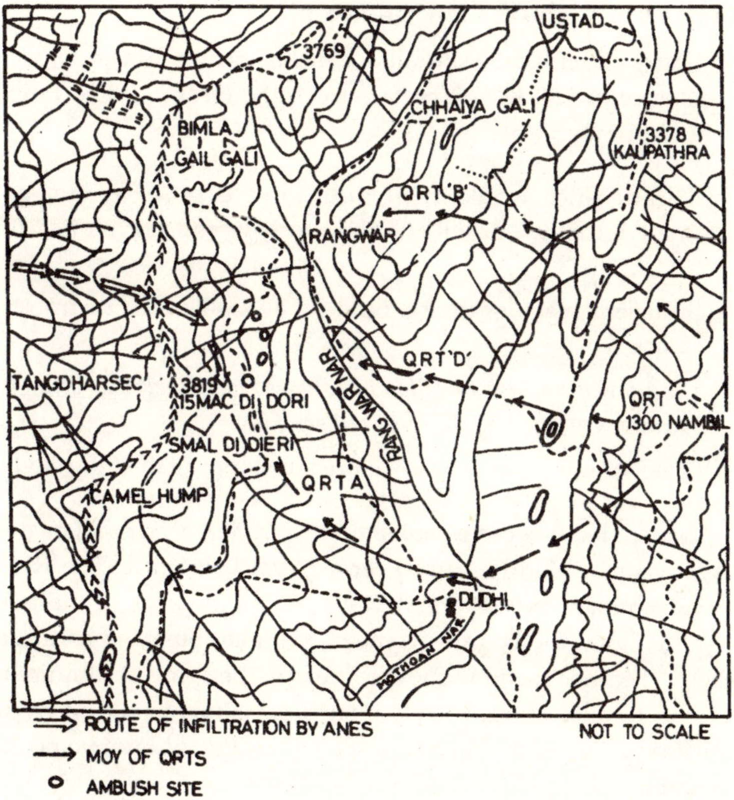

noted and the gaps identified. The guides would then position themselves at every suitable gap and after last light, guide the group to their location by means of a torch. The waiting groups would then climb towards the light, and after resting for a while and orienting themselves, cautiously work their way between and behind our ambush positions.

Genesis of Ambush Plan 'Dhanush'

It was decided to site all ambushes on the lower slopes, at a distance of 500 to 700 metres from the crestline. The rational behind this step was:

- The distance would help to ensure surprise, as any noise inadvertently caused by our troops would not be heard by the ANEs.
- Sufficient reaction time would be available for our troops to react.
- *The infiltrating ANEs would not be able to work their way behind our positions as had been the case hithertofore.*
- Observation was relatively easier upslope, against the skyline.

Patrol bases were established in pup tents. Surveillance was organised by buddy pairs keeping watch in turn. The RL fire using ambient moonlight/starlight, without prior illumination, would signal the commencement of the ambush, so as to deny the ANEs, the reaction time afforded by the firing of illumination round, to take cover. To retain surprise, stops would move into position after the ambush was sprung and would not be position ab initio.

At about 2015 hours the scout party commander noticed move on the skyline. He immediately informed the ambush party commander. The militant group crossed the range between Ismail di Dori and Gail Gali. The group was about 40 strong and were moving on a line which would bypass the killing area of ambush by about 400 metres. The Company

Commander had to make a quick decision, either to move the ambush site forward, or to engage them. He decided to engage them. The firing of the RL was the signal for the ambush to open fire and all weapons engaged the ANEs. For at least two minutes the ANEs did not react as they were totally dumbfounded. Unfortunately, a stoppage occurred in the MMG as one fired case had got stuck in the chamber. The gun was put back into action after the stuck fired case was pulled free with a bayonet. After the initial volley office, the ambush party commander ordered the various weapons detachment commanders to exercise fire control and indicated the targets to be engaged by them. Immediately after the commencement of the action, the ambush party commander contacted the platoon commander at Gail Gali and ordered him to move to establish a stop position at point 3721. The CO, on receipt of this information, ordered five QRTs for establishing blocking positions to prevent the escape of the militants. While searching the area these parties shot and killed three ANEs who were trying to escape back to POK (Pakistan Occupied Kashmir). The surviving ANEs had managed to disperse and using the folds of the ground made their way towards the crest of the Shamshabari in ones and twos, turning around to snipe at the pursuing troops.

The search progressed well into the next day when two wounded ANEs were captured from where they had taken refuge in the cover of boulders. The following weapons/ equipment were recovered from the killing ground:

Rifles AK-47 (T-56 Type)	- 28
UMG	- 01
Sniper rifles (Dragunov)	- 02
Pistols	- 19
Telescopic sights	- 03
Magazine (AK)	- 63
Magazines (Dragunov)	- 03
Magazines (Pistol)	- 24
Ammunition (AK-47) (rounds)	- 2800

Ammunition (UMG) (rounds)	- 450
Ammunition (Pistol) (rounds)	- 363
Radio sets (Icon)	- 7
Walkie Talkie sets	- 2
Hand grenades	- 25
Electric detonators	- 137
Detonators (No 27 type)	- 96
Timepencil switches	- 37
Remote decoder device for radio controlled explosive devices	- 6
Currency Indian Rupees	Rs. 2,700/-[1]

With latest knowledge, it appears that there were three groups of 15 each belonging to Karwan-e-Haider, Al Jehad and HM. In addition there were 5 guides. This action when viewed against a backdrop of high altitude, severe cold, inhospitable terrain, high snow and other negative conditions speaks volumes for the Company Commander's dedication, devotion and above all the motivation in the men, who despite all odds, gave a fine account of themselves against a militant group twice their size.[1]

An ambush took placed on an Army convoy of four vehicles in Manipur on April 10, 1991. This was deliberately planned and clinically executed by the NSCN by combining the effects of careful selection of the site, use of IEDs and intense firepower. On the part of the army the party was not able to manoeuvre itself in time to save the weapons of the JCO and three other ranks in the first vehicle which obviously took the brunt of the ambush. Here is the case study of the ambush, which gives the details of the effective NSCN ambush.

OPERATION KACHAI

General

OPERATION KACHAI is a case study of a hostiles' ambush

on the CO's convoy of Battalion 'A' which was deployed in Western Ukhrul district of Manipur, the hotbed of Tangkhul activity. The dastardly operation resulted in the death of one JCO and three other ranks including loss of arms.

It is a classic operation which highlights the need to carry out an introspection about our vigilance drills, and adherence to laid down SOPs (Standard Operating Procedures) besides a variety of other factors dwelt upon in this case study.

Background

Ukhrul district is traditionally a bastion of the National Socialist Council of Nagaland in this district since August 89. NSCN had suffered a number of setbacks in chance encounters with the battalion. On January 3, 1990, the battalion encounter with NSCN at Upper Tingsong in the resulting death of two undergrounds, recovery of 97 rounds of 22 ammunition and other assorted material. On December 10, 1990, in another encounter Sgt Mario was killed and SS Cpt Shanghai was captured. On February 26, 1991, two UG (underground) girls were apprehended who were carrying important letters for various NSCN department heads! All these actions took place in the area North and North-West of Kachai, which was one of the posts of the Battalion. These setbacks resulted in the loss of NSCN credibility among the local population. To re-establish their lost credibility, as also to dominate the impending mid-term elections, the NSCN had increased its activities in the area of responsibility, in particular around Kachai, during the first four months of 1991.

On night April 6/7, 1991, at about 2345 hours, the CO received a radio telephony message from Kachai that the suspicious move of two persons had been observed by one of the sentries through a night vision device near the outer fencing of the post. Before the sentry could open fire, they were seen running away from the post. The Post Commander was told to send a QRT around the post and also carry out a selective search of the village. Sensing the gravity of the situation, the CO left for Kachai and other posts

OP DHANUSH 7/8/ DEC 92

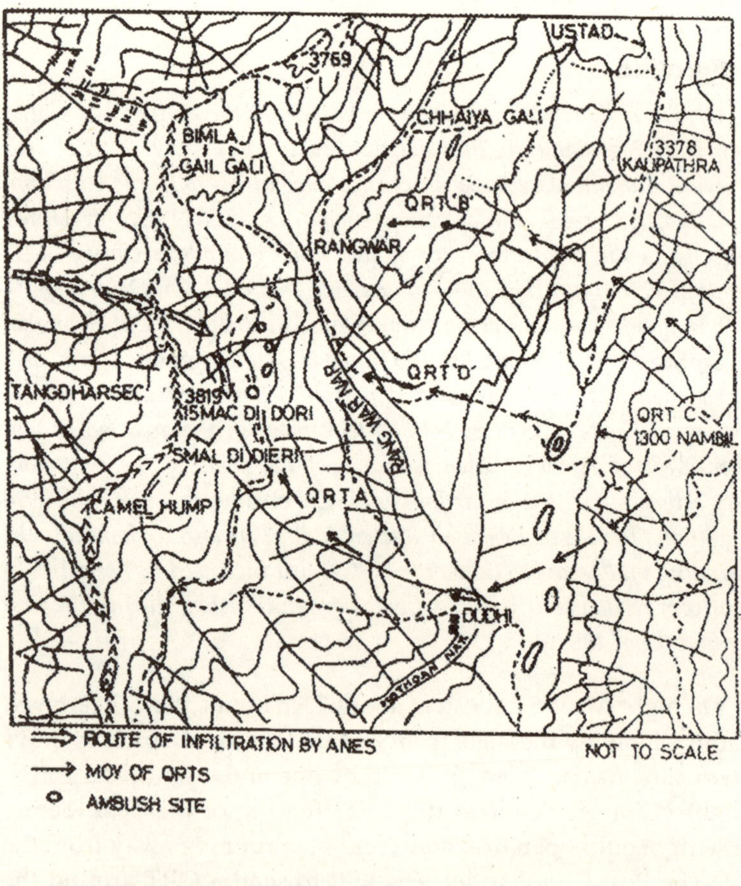

in Iril River valley on April 7, 1991. He stayed at each post for a night, briefed the post Commander with regard to the conduct of our own operations, NSCN activities and their likelfy intentions. On April 10,1991, he left Tora at 0830 hours for Battalion HQ via Kachai. He had a short halt at Kachai post and left for Battalion HQ at 1010 hours. There was no road opening party on the way since it was not possible to cover the complete road stretch from Tora to Battalion HQ (Tolloi) due to the paucity of troops, as the troops from Lower Leishan and Tora post were involved in shifting of Lower Leishan post to Tora. Kachai post had provided a patrol of 1 JCO and 20 other ranks in between Tora and Kachai.

Ambush

There were four jongas in the convoy with their tarpaulins rolled up and the CO's vehicle was second in the order of march. When the convoy reached the road bend (marked on the sketch) and the leading vehicle was negotiating the turn, there was a sudden explosion accompanied by intense automatic fire from the direction of the road bend and hillside on the right. The other there vehicles of the convoy which were moving at a distance of 30-35 metres apart halted and men started firing towards the source of opposing fire. The CO alongwith two persons rushed towards the leading vehicle and meanwhile, Subedar Major with two men climbed the hillside and covered the moved of the CO's party. The CO also ordered others to follow him to carry out the pursuit of fleeing UGs. On reaching the location of the leading vehicle, it was found that one JCO and three other ranks were dead, there Sten Machine Carbines and two SLRs were missing and two other ranks laying next to the Jonga were injured. Meanwhile the remaining protection party alongwith the tracker dog fetched up. A pursuit party of one havildar and 10 other ranks alongwith a tracker dog was sent in pursuit of the fleeing UG.[2]

Notes

1. *Prativirodhi,* March 1994.
2. *Prativirodhi,* April 1992.

Flexibility

An army may be compared with water, for water in its natural flowing avoids the heights and hastens downwards so in a war, an army should avoid strength and strike at weakness. As water shapes its flow in accordance with the nature of the ground, an army manages to be victorious in relation to the enemy it is facing. As water retains no constant shape, so in war there are no constant conditions. One who can modify his tactics with the enemy's situations and succeed in gaining victory may be called a divine.

Flexibility of the mind is the basis of this principle of LIC. This principle of LIC is an important factor of insurgent tactics, who are not bound by extensive logistic back-up systems, posts, fixed routines, areas and uniforms. Fluidity of operations makes their existence like the breeze, which are easily felt, yet difficult to hold.

This degree of flexibility may not be possible for the army but flexibility combined with initiative can certainly blunt the tactics of the insurgents. Flexibility will only yield results when a commander is willing to take initiative and change or modify his plans to either surprise the enemy or to react to an unexpected situation that may have developed during the course of the operations.

Flexibility of mind is also a prerequisite for changing basic policies, like acceptance of surrenders and grant of general pardon to all insurgents who surrender willingly. This was successfully attempted with the ULFA and resulted in large-scale surrenders of the low-level ULFA cadres, thus breaking grass root-level support of the organization.

Besides initiative, mobility and Communications are the other principles which are prerequisites for a flexible approach. Mobility and speed are required to change the location of troops and switch them to the required area of operations. This need not relate specifically to vehicular movement, as mountain-based insurgency requires extensive foot movement to achieve flexibility.

The passage of orders for switching of troops has to be quick and accurate. This can only be done by a reliable and effective means of communication. Besides being reliable it is also essential that communication is secure to avoid false transmissions on the net. Another basic requirement for achieving flexibility is decentralization of powers and the options for reacting to varying situations. If a sub-unit commander has to look over his shoulder for detailed orders, flexibility will remain a principle which will not be converted to a battle-winning formula.

Flexibility can also be achieved by having reserves or, in terms of LIC, QRT. This team provides the option of moving troops at short notice from the location of the force to the specific area of operations. It can be compared with the light cavalry regiments of yesteryears which were thrown at the enemy to break the design of his battle.

The location of these QRTs is as important as their availability. These teams should be located in such a manner that they can react to a variety of situations, without having to traverse long distances. These teams should also be provided reaction with small, light and fast-moving vehicles to ensure fast reaction.

Coming to the company level the following actions can help achieve flexibility during LICs.

The JCOs and NCOs should be trained to undertake independent actions. Unless junior leaders have confidence in themselves to perform without orders at each stage, flexibility will not be possible at the company level. They should be a part of the intelligence gathering process, planning and briefing in the company to enable them to be in the picture of the situation and react accordingly. Mistakes will have to be corrected and taken in the stride, otherwise they may lead to the curbing of the initiative of the junior leaders.

Operations should preferably be planned with small teams, which are capable of moving fast and changing their positions quickly. These teams should be led by leaders used to independent action and freedom of thought.

A high level of physical fitness has to be ensured in the company to enable the troops to cover larger areas in as small a time frame as possible.

Good knowledge of the area of operations is also important as taking cross-country routes both during day and night is only possible when each and every area is familiar to the teams. This has however to be reinforced by reliable local guides.

Feints and false movements are another difficult but effective means of achieving flexibility. This can overcome the paucity of troops and it should be coordinated with the battalion HQ to ensure coordination.

A peculiar yet striking example of flexibility came to light during the German invasion of the USSR during the Second World War. After the Germans had inflicted telling losses on the Soviet Army, they did not pay adequate attention to the dispersed units of their Army. A large number of soldiers refused to surrender and exchanged their military uniforms for civilian clothing and sought refuge in villages, and others hid in swamps and forests: the more active of them organized guerrilla bands.

These guerrilla bands played havoc with the German supply lines, communications and small garrisons. These bands varied from 15 to 200 in strength depending on the terrain, the availability of volunteers, and the attitude of the populace towards the German invaders.

Greatly alarmed by the new threat, the German High Command initiated intensive study of appropriate methods of suppressing the guerrilla activity. To accomplish this each CO was authorised to do what he deemed best. Certain of them tried to hunt down the guerrilla bands by despatching combat units into the woods. This method proved to be unsuccessful. If not confronted by superior guerrilla forces or decimated in a trap, the companies or battalions returned after an abortive search with empty hands. The great advantage of guerrilla bands in the jungles lay in their mobility and their ability to disperse and disappear amongst the population, their excellent communication with local inhabitants, and in the clumsiness of the regular army units inexperienced in backwoods fighting. Efforts to eliminate the guerrilla movement through retaliatory measures against the population were even more disastrous.

In spite of this hopeless situation some German commanders did find ways and means to deal with the supposedly helpless situation. The flexible means adopted by them helped them eliminate the threat of the guerrillas, or at least reduce it to the bare minimum.

The most outstanding method was adopted by the cunning Gen of the German Second Panzer Army, Gen oberst Col Gen Schmidt. In early December 1941 this army had been stopped south of Moscow and had fallen back to the line Zhizha-orel-Kursk. To their horror they found that except for big towns like Bryansk,

Bezhitsa, Karachev, Donitrovsk, Dinitriev and Sersk all areas were under guerrilla control. The supply lines for the complete army constituted a single railroad, Orel-Karachev-Briyansk-Unetcha-Orsha-Minsk and the highway Orel-Briyansk-Roslavi-Smoleusk. The other areas around Briyansk, including Lokot and Trubchevsk and north of this city, including Dyakovo and Lyndinovo, were controlled by guerrillas. In the spring of 1942 the towns of Dyatkovo and Lyndinovo were cleared of the guerrillas by the Germans. However their garrisons were connected to one another by means of one road and a branch railroad line which was threatened by guerrilla attacks.

The strategy adopted by Gen Schmidt was ingenious. In March 1942, a horse-drawn sled from Karachev arrived at Navlya and Lokot, bringing a small group of Russian civilians. The chief, one Kiminski, a slender, energetic, middle-aged man, presented to the German garrison commands "to whom it may concern" orders signed by Gen Schmidt which requested German units to render every possible assistance to the bearer of the order. Furthermore, the order appointed Kiminski as Governor of the area including the towns of Navlya, Lokot, Dmitrovsk, Dimtriev and Sevsk. He was authorised to act independently to appoint local officials, to organize the economy of the area, and, what was more important he was responsible only to General Schmidt.

The new governor immediately appointed mayors, proclaimed the abolishment of the collective farming system, supervised the destruction of the remaining insurgents and started the organization, implements and stock among farmers. From then on every cow, horse, pig and cattle was their private property. Owners turned their wrath against guerrillas who soon became a formidable brigade, consisting of five to six battalions of 500 to 600 men each, a tank unit with 10 or 12 light tanks and an artillery battalion with some 20 guns. This idea followed the principle of flexibility and was able to do what regular German troops had failed to achieve. A more recent example of the employment of the principle is amply illustrated by this case study of army action against the insurgents in the Kupwara sector of Jammu & Kashmir.

OP Jackal

Dardpura is a remote village on the upper reaches of Hamhu feature in Kupwara district. It is at an altitude of 7000 ft and surrounded on three sides by mountains having steep slopes by virtue of its location. The complete area for miles is under observation thus making it very difficult for the security forces to move in with surprise. There are innumerable nalas thereby providing plenty of escape routes. Most of them flow from West to East and finally merge into the Hodkol.

Late in 1993 it was noticed that all sources working for this unit were being killed one after another. Investigations revealed that this was the handiwork of only one militant, Mangta Lone (Codenamed — "Shapo") of Dardpura village. He belonged to the Al Barq group and was brutally murdering the sources after torturing them.

It was decided to nab him. Somewhere along the line, he became a District Commander, and let loose a reign of terror. Sources clamped up; some even refused to whisper his name.

Attempts by various units to nab him yielded no results. At night he would change his spots more often than a leopard; during the day he was the elusive shadow. By his escapades he become a legendary 'figure and acquired an aura of invincibility. Thus the more he got away, the more determined we were to get him.

A source revealed that Shapo had visited his house for two hours on July 31 and was likely to repeat it the next day.

The Company Commander immediately swung into action

and hatched an ingenious plan. The officer knew that if Shapo even had a shadow of a doubt the operation would be off.

The plan involved occupying at midnight a prenominated house next to Shapo's. The group would lie doggo, and if and when Shapo turned up, shoot him.

On midnight July 31/August 01 the house was occupied. A plank in one of the rooms was pulled open and then replaced so that it could form a loophole in an emergency. At 0720 hours the spotter informed the officer that the man who had just come and was sitting outside was none other than Shapo.

The officer asked him to surrender. Shapo replied by the burst of an AK rifle. As pre-arranged, NCO sniper was to fire one round from his sniper rifle, which he did. The bullet found its mark on Shapo's neck. There were two more militants who could not be seen. They immediately opened up. The QRT charged out of the house, but in the meanwhile these militants mingled with the population. Taking them on would have resulted in casualties to innocent civilians. Tremendous restraint was exercised. A quick search of the area revealed the following:

District Commander Al Barq "Shapo"	Killed
AK rifle	3
AK magazine	6
Ammunition (rounds)	1900
Hand grenades	12
60 mm mortar bombs	3
107 mm arty (round)	I

Note

1. *Prativirodhi,* March 1994.

Initiative

Losing the initiative means to be defeated, to be annihilated.

— Mao Zedong

The principle of Initiative has already been seen as the guiding light to other principles such as Surprise and Flexibility. Initiative has been defined as, *"take the first step on one's own without an order or suggestion from others, have the right or power to take the first step; show ability to start new schemes."*[1] This definition aptly brings out the meaning and spirit behind this word. It symbolises action, speed, activity, decisions, movement, daring and, finally, victory.

I am reminded of a story of Akbar and Birbal. Akbar as usual came up with a riddle and said, "Who can place an apple on a slanted table top?" Everyone tried to place the apple but as soon as they would leave it, the apple would roll down. Ultimately disappointed Akbar asked Birbal if he could do it. Birbal got up, walked to the table, placed the apple on it and pressed it hard. The apple stayed put on the table. A courtier said, "If this was the way, even we could have done it." Birbal smiled and replied, that it is the initiative of the first man to do a task, thereafter anyone can do it. The answer was accepted by Akbar.

Initiative stems from taking bold and calculated risks which will take the insurgents by surprise. It involves taking action without formal orders from one's superiors. This principle can only be employed if the commander has self-confidence, thorough knowledge about the ground, accurate intelligence about the insurgents and confidence in his command to carry out his instructions and make his decision worthwhile.

Initiative also implies looking for better ways to solve a problem or work at a situation. It involves the ability and will to experiment and try new ideas to solve a tricky riddle. But again experimentation should be based an hard facts and not figments of the commander's imagination. This was illustrated by the US Army fighting counter-insurgent operations in the Philippines at the turn of the century. After Aquinaldo's army was defeated and Aquinaldo himself captured some of the extremists took to the hills and became guerrillas. They roamed the jungles raiding, ambushing, killing and pillaging. Finally the United States found the solution to the guerrilla problem in the Philippines itself. They recruited native Filippinos-men wise to the jungle ways, men who knew the trails and mountains like their own backyard. They were divided into small groups of 10,15, 20 or 50 men and over each group a trained American officer was put. The trick was constant patrolling over each trail, and careful attention to intelligence work. The jungle, night time, surprise attack and initiative are the guerilla's weapons. The solution is to adopt the same weapons to fight him.

Initiative also involves planning well ahead to remain one step ahead of the insurgents. This will mean that the commander must have the ability to carefully predict the actions of the insurgents based on the intelligence inputs he would have received.

This principle of LIC should however not be used against other posts and paramilitary forces, to get to an objective that others have been planning for in consultation. This trait of one-upmanship will not only destroy the credibility of the company, but it will cause strikes before the time is ripe for these.

At the company level the following actions can be initiated to achieve desirable results from this principle of LIC.

A Company Commander must act on reliable information rather than waiting indefinately for the ideal information. Situations will have to be changed to suit one's requirement rather than waiting for set-piece action.

Tasks which involve risk and danger but promise good chances of success must be seized. A soldier will very much prefer to risk his life when he is going in for the kill, rather than routine forced operations which rarely yield results.

Once contact has been established, one must never lose initiative and in turn lose the insurgents. A chase must be given and the insurgents should be killed or captured. A little initiative and a little extra effort will completely change the complexion of a situation.

Initiative should also be taken to undertake meaningful civic action projects and developmental work at the company level.

Lastly the Company Commander must initiate meetings with the people and encourage the flow of information as this will remove the mental block in the minds of the people, and make the army more approachable.

Sun Tzu attached great importance to gaining the initiative in any confrontation. His principles of taking pre-emptive measures of fighting a quick battle to force quick decisions, bringing the enemy to the battlefield, instead of being brought there by him, are all for the sake of gaining the initiative in war.

The Liaoxi-Shenyang Campaign during the Chinese War of Liberation (1945-49) is an example of this tenet. During this campaign, the tactical situation was not favourable to the PLA (People's Liberation Army). Despite having beseiged Chang Chun and crossing Songhua Jiang River three times they failed to seize the initiative. Lin Biao's army was superior to that of Wei

Linhuang's, the Kuo Mintang (KM) Gen. However, Mao Zedong and Zhou En Lai of the PLA felt that this was due to the mistakes of the PLA commander.

After deep thought, the two generals, Mao Zedong and Zhou En Lai decided that they should attack Jingzhou, a strategic decision whereby the enemy would be forced to defend an important location. Jingzhou served as a key location between Northern and North-Eastern China for supplies. This would be taken, supplies for the KMT Army in the North-Eastern areas would get cut off. Besides, mutual strategic support would be lost. After Jingzhou was attacked the enemy would be forced to scamper to its rescue in a strategically weak position, due to the PLA's initiative. If the place was reinforced, the KMT Army would be risking getting it wiped off, while they were on the move, thus creating the condition of flexbility for the PLA.

The decision taken by Mao Zedong and Zhu En Lai was proved to be correct. As the two had expected, while Jingzhou was being attacked by the PLA, KMT sent reinforcements from Northern China which were intercepted by the PLA and failed to reach Jingzhou. KMT troops were also despatched from Shenyang to rescue Jingzhou but were intercepted and annihilated at Dahushan. Later the KMT Army surrendered. In this campaign more than 470,000 KMT troops were wiped out. The whole of the North-Eastern region was liberated.

It can be concluded that strategic initiative is not innate in a battle, but has to be created by wise commanders. Flexibility of tactics in a campaign is generally obtained on the basis of strategic initiative. The Liaoxi-Shenyang campaign illustrates the truth of this theory.

The principle of initiative was also effectively used during OP ADVENTURE PATROL, KATAR in Doda District of Jammu & Kashmir.

OP ADVENTURE PATROL KATAR

General

The general area Dhanaso, Dhamot Di Gali and Bagar was a conduit and hideout for move of ANEs from general area Pingle, Champal to Bharthe and Bhatmas. Based on information of such move of ANEs, an adventure patrol was launched by 'C' Company, on August 18, to August 20, 1994 with an aim to seek an encounter.

Troops

One platoon 'C' Company, composed as under:-

(a) Officer	-	*1 (Capt Rajesh GK)*
(b) JCOs	-	*2*
(c) OR	-	*38 (including MMG detachment, Engineer representative and Nursing Assistant).*

Outline Plan

Move along ridge line Khakri, Dhamot Gali, Suin ki Gali and further to Bhaiya, Jugassar and back.

Gain real time information about ANEs en route from locals.

Seek encounter with ANEs based on information provided by locals.

Execution

The adventure patrol left Guno at 1400 hours on August 18 and halted at Dhamot Di Gali for the night.

The patrol left Dhamot Gali at 0800 hours. En route it met a few Gujjars, and Capt Rajesh GK casually spoke to them. Information about the presence of ANEs ahead was revealed by the Gujjars. The officer appreciated them to be located on a particular ridge line. The patrol was momentarily stopped to brief troops to increase their vigilance

OP ADVENTURE PTL KATAR

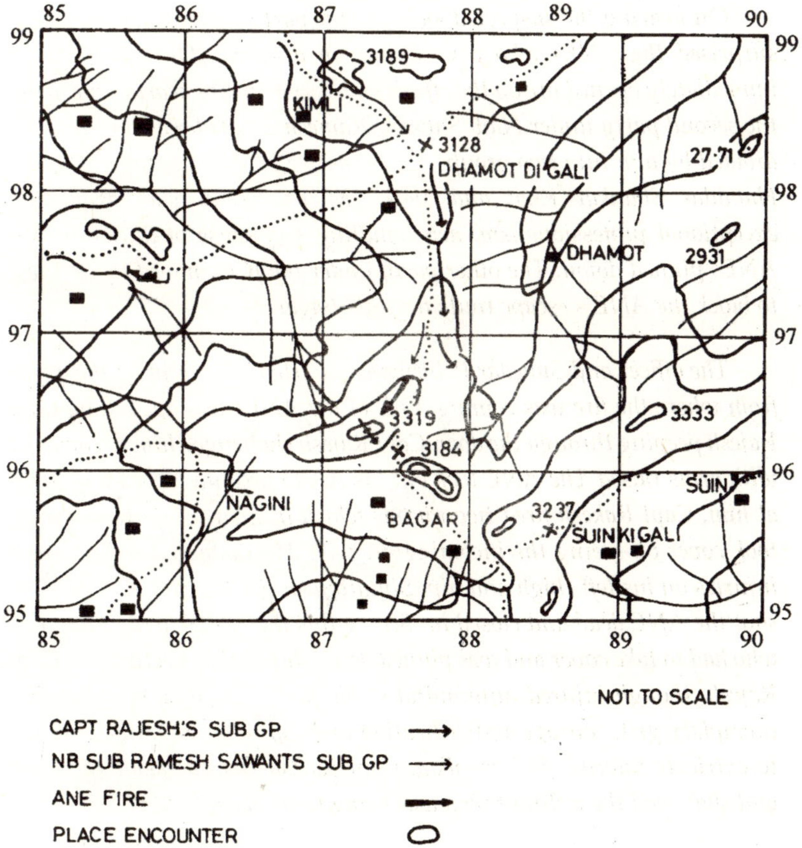

and divided into two groups. One party under Naib Subedar Ramesh Sawant alongwith the MMG Detachment approached Point 3184 from North along the ridgeline and Point 3319. The other party under Capt Rajesh himself approached Point 3184 along Guglala Nala from the East, taking a very difficult and covered approach to the suspected area.

On nearing the suspected location, the party under Captain Rajesh surprised the ANEs who were hiding in a gwari (Hut). The ANEs immediately opened fire on the officer and his party. On hearing the firing the second party under Naib Subedar Ramesh Sawant was immediately divided by him into two groups. One was the MMG detachment under Havildar Shankar Khot and Naik SM Kumbhar, who, displaying exceptional professionalism, deft handling and sense of duty kept the ANEs pinned down. The other party under the JCO moved to the West to block the ANE's escape route towards Bagar.

The officer and Subedar P Tarihalkar started closing in on the gwari from where the fire was coming. One of the ANEs was spotted by Capt Rajesh peeping through the door. Capt Rajesh challenged him to come out with arms raised The ANE took out his AK to fire and lobbed a grenade at him. Capt Rakesh fired two rounds which injured the ANE and then took cover on seeing the incoming grenade. He sustained four splinter injuries on his left thigh and hip. Simultaneously Subedar P Tarihalkar shot the ANE dead and closed in. Seeing this the ANEs fired at the JCO who had to take cover and was pinned down due to the effective fire. Capt Rajesh, though injured, unmindful of his personal safety and showing exemplary grit, courage and dedication to duty, closed in onto the gwari to extricate Subedar P Tarihalkar. Capt Rajesh rushed inside the room and shot dead the militant who was firing from inside.

As the number of militants present in the adjacent room was not known, Capt Rajesh moved back and ordered the fire of 84 mm RL-HE (High Explosive) on the adjacent room. Sepoy Gawade Bapu and Sepoy Awatirak SK showed exemplary courage by closing in and lobbing grenades to ensure the elimination of all the militants inside. However, due to a mud wall the ANEs were still hiding and kept on firing to prevent anybody entering the gwari.

Effective grenade lobbing by Sepoy Gawade and Sepoy Awatirak forced one ANE to escape through the window. But Havildar Khot and Naik Kumbhar shot him dead thereby preventing him from making good his escape.

Capt Rajesh again moved into the adjacent room and found one more militant, though injured, firing through a window and shot him dead.

Results Achieved

(a)	*ANEs killed (One foreign mercenary)*	*-5*
(b)	*Weapon;*	
	(i) AK-56	*-3*
	(ii) AK-47 (with wooden butt and encrusted with jewels 'HM').	*- 1*
(c)	*Magazines*	*-10*
(d)	*Ammunition (assorted)*	*-139*
(e)	*Fired cases*	*- 62*
(J)	*Radio set*	*-1*
(g)	*Binoculars*	*- 1*
(h)	*Grenade*	*- 1*
(j)	*Audio cassettes*	*-3*
(k)	*Diary and incriminating documents.*	
(l)	*Other warlike stores.*	

Note

1. Homby, A.S., *Progressive English Dictionary.*

Security

The success of an operation depends on the principle of Security. This vital principle of LIC makes the difference between honour and shame. It is the endeavour of the insurgents to gather intelligence about the army, and gain the initiative.

A vast variety of means are available to him to find details about the activity of the army. Amongst these are sympathetic locals, informers, tell-tale signs and radio transmissions. The protracted character of operations sets in monotony, which in turn breeds boredom and casualness, the basis of security leaks.

There are a number of indications of impending operations for the insurgents. It is these signs which indicate the move of the army. Increase in radio transmission is one factor which is bound to occur as the to and fro passage of instructions will accompany the operation. The fundamental principles of radio procedure, short transmission and cutting down of elaborate orders come into play. The standards of battle procedures also go a long way in reducing transmissions.

The move of the company in vehicles or on foot in a particular direction sends the insurgent radio sets crackling. Move on foot can be carried out at night to ensure security of the move. In case the move has to be carried out during the day, circuitous routes can be restorted to, or move of more than a single patrol in a number of directions can help confuse the insurgents. In case of vehicular movement at night the vehicles can continue to move along the road and the Company can dismount from the moving vehicles. These are just a few ways where the situation and specific characteristics prevalent will dictate the employment of the means to achieve security.

Peculiar yet common problems like barking of dogs is an unavoidable evil if a village happens to fall in the area of operations. It can however be ensured that those affected areas are not crossed en route to the area of operations. Bypassing them will prove to be a more secure operation.

Loose talk will always be an evil which will sabotage operations and should be guarded against. In their enthusiasm to show their knowledge and participation in major operations, at times troops tend to talk about them. The principle of "need to know" should strictly be adhered to, to avoid such situations.

In the LIC environment it is desirable that troops be regularly educated about this aspect. The aspect of security should be dwelt upon 'in sainik sammelans' and informal gatherings in the Company.

In the Indian LIC environment, security of posts has not been a maj or threat. However stray incidents like the 'Oinam' post raid in the north-east and jitter actions in the valley have to be taken in view to plan the security of the post. Alarm systems, orders for move of QRTs and the information network should all be disseminated to the company. These actions should be rehearsed to ensure their feasibility and practicality.

Security of the Company personnel is another important task. The success or failure of their tasks in LIC is gauged by the number of kills or the number of casualties. It thus becomes the personal responsibility of the Company Commander to ensure this aspect. The methodology of ensuring it may vary, but in an unconventional battle results are what matters. This is how Bob

Sanders, who was a young Private in Vietnam felt about his Company Commander.

"Now that I look back upon it, the Company Commander was doing his job. But he used to push us, man, until our bones ached. We would be wanting to stop and rest, stop and cook and eat and he'd just push us, some days he'd push from can't see in the morning to can't see in the evening and we'd be cursing and passing at him. But I give him thanks, too, because if it was not for him driving us on in such a way, I might be dead today. We called him Rabbit, he was the best CO we had in Nam and 1 had two others besides him. 1 never knew his real name. We called him Rabbit because he wasn't like the average CO walking through the jungle; he'd be running through the damn jungle. But he was good, really sharp. Half the time, we wanted to kill the bastard and the other half of the time we loved him because he was that good. And he was fair.

You gotta have a fair CO. You got these chicken shit COs that hadn't been through any real combat. They'd be straight out of O.C.S with all this John Wayne bullshit. They'd come over and want to read out of the book on how it should go, man. You know standard operating procedure. And their book would get you killed. When Rabbit was there we dumped the book you could not go by the book in Vietnam. Maybe in previous wars you could say, "We hide behind this, or we move over here and this and that." But when you got a good company commander he used his own discretion in certain situations. Lotta times the shit hit so fast the book didn't help. It was constantly up to you to react in a certain way. If you reacted wrong you were dead.

Take the Lshaped ambush. When you fell into that you knew right off where the major firepower was coming from. If you had a shit company commander, he would take just one glance and what was happening. If you had one that had his face buried down in the ground, soon as the shit started coming in he was trying to hide from the heat then your whole unit would get killed right there. Rabbit never got down on the ground.

Half the time he was the only one up checking things out. Where was the automatic weapons fire coming from? Where was the main body of the ambush? Then he would give you orders to move in the right direction. We knew 95 percent of the people in the unit had total confidence in him; we knew that he could get us out of trouble. That confidence was very important. You had to remember that the company commander was only human. He could be out there for some bullshit purpose. As a captain he was a career officer; he wanted to make rank. I've heard that there were company commanders in Vietnam that actually got slugged. They didn't give a damn about their men. When nightfall came and nobody could see, some grunt would take a hand grenade and just blow him away. It was that simple; hand grenades don't leave finger prints.

We had a lot of confidence in Rabbit. He was like the Godfather. He was everything to us. His direction was our destination. Now just cause he was a captain, people maybe felt they had to take orders from him. We didn't have to take orders from him. We didn't have to take orders from anyone. At least I felt that way. He had a weapon. I had one too. We loved that guy and when he left us in the field, he made a little speech, that was supposed to get us ready for the next company commander. He told us to drive on and stay strong, and remember the spirit of the people could move mountains. It was given in a way that affected us. We held on. Some of the guys are dead and gone and I loved 'em all. He tried to instill in us that there wasn't no black and white in Nam. You forgot about the shit. I felt that way anyway.[1]

The lack of security can be disastrous, as was proved by an NSCN ambush on an army convoy on the Sita Tengnoupal road. The company learnt the lesson of principle of security but the hard way.

OP HIFAZAT

Introduction

On 29 June 1993 an Army convoy of 1 officer, 4 JCOs and 66 other ranks was ambushed, while moving from Sita to Bishanpur in Manipur, approximately seven kms from Sita (a post) in which the Company suffered heavy casualties and lost arms, ammunition and equipment also.

Events Leading to Ambush

The unit, on induction, was deployed in a dormant sector. Since the sector was comparatively peaceful, units/sub-units were moved frequently for operations to sectors which were more active. During early 1993 Naga UGs started choking Moreh township by laying ambushes on civil traffic, resulting in the virtual closure of NH (National Highway) 39. The unit was tasked to open NH 39 which it did with great zeal and determination, and kept it safe and secure for civil traffic.

In the month of March 1993, a large number of villages were burnt which led to migration of refugees to safer and larger villages for security reasons. A large number of Kukis concentrated at Sita and Tengnoupal. In order to prevent inter-clan clashes between the warring tribes, the unit kept constant pressure on Naga UGS and launched several successful operations. Unfortunately, the casualties and pressure on the Nagas led to malicious campaigns against the unit in the regional/vernacular media. Consequently the CO received some threatening calls and letters from Naga hostiles.

On June 28, 1993 the company which was located at Sita was ordered to move to a location at Bishanpur and subsequently both companies were to move for operations. At approximately 1700 hours these five vehicles left for Sita. Details of this move were passed by the Adjutant to the Company Commander in clear on the radio set. At approximately 1945 hours the CO also spoke to the Company Commander in clear and briefed him regarding the impending move and instructed him to reach Bishanpur by 1200 hours on June 29, 1993.

After loading the vehicles, the company commenced its move

from Sita at approximately 0730 hours on June 29. Since it was declared to be an operational move no ROPs were considered feasible and necessary due to other commitments. In any case companies had been moving without ROPs on several occasions in the past due to the dormancy of the sector. The company had identified certain areas as ambush-prone areas and was expected to cover the stretch of road on foot. However, due to paucity of time, as the Company Commander was expected to reach Bishanpur by 1200 hours, this practice couldn't be followed.

Ambush

After traversing a few kms the vehicles had spread over a large distance due to a bad stretch of road. The Company Commander stopped and reorganized the convoy. However when the convoy restarted, vehicles had closed in and the distance between the vehicles reduced to less than 100 metres.

At approximately 0830 hours when the vehicles reached Km stone 7 or, road Sita-Tengnoupal all vehicles came under intense automatic fire from the hillside on the right side of the road near Chelep village. The driver and co-driver of the leading vehicle probably got killed in the first volley of fire and the vehicles stopped immediately, thereby blocking the road and preventing other vehicles from escaping.

Other vehicles were also engaged simultaneously and stopped one behind the other over a distance of 175 metres, inside the killing ground of the ambush. Fire was very intense and accurate due to the proximity of the vehicles with the hostiles' firing position, which at best would be 7 to 10 metres.

Reaction by troops

The company suffered maximum casualties in the first volley of fire as a large number of persons could not even get out of the vehicles: those who dared to jump from the vehicles were felled close by. After jumping from the vehicles some persons took positions under the vehicle and attempted to engage the insurgents. From these positions they possibly could neither see the hostiles nor engage them. 2" mortars could not be fired as ammunition and mortars were left in the vehicle while jumping.

Persons who had taken position on the road, under the vehicles and along the vertical cutting found themselves exposed to grenade splinters of the UGs and therefore were forced to abandon their positions.

Efforts to communicate with neighbouring posts proved futile as all the radio sets were out of action as the radio operators while jumping could not take out the radio sets.

Unfortunately all but one of the radio sets were kept in the single vehicle of Company HQ which had suffered the maximum firing of the hostiles.[2]

Notes

1. Malone, Col Dandridge M., *Small Unit Leadership.*
2. *Prativirodhi,* March 1994.

Intelligence

Know the enemy and know yourself.

— Sun Tzu

Intelligence forms a very important principle of LIC and this principle alone can make the difference between unsuccessful, frustrating "jungle bashing" operations and operations based on specific information which led to encounters and results. Intelligence is the principle which takes the maximum time to collect and collate after move to insurgency-prone areas. Unlike what it is perceived to be, intelligence is an entire subject in itself which requires systematic management to achieve the desired results.

As has been discussed earlier, Intelligence can be classified as generic and specific or actionable. Whereas generic intelligence forms the basic framework, specific intelligence gives it the direction for operations to be launched.

The role of intelligence agencies has often been the subject of debate in insurgency. The agencies have been accused of feeding generic information alone whereas specific and actionable information has been lacking. Seeing the basic structure of the intelligence agencies and their role, it is quite difficult to see them collecting actionable information; though they certainly are suited to give background intelligence. A Company Commander must understand and have no expectations in this regard. It is he who will have to work on the background information provided to extract specific information. Frank Kitson compares these intelligence agencies with cows which can turn grass into milk but the further conversion of milk into butter has to be done by us: the Company Commanders.

Generic information should be collected to give a broad pattern of the working of insurgents. It can give an insight into their morale. This information will give guidance for increasing the pressure on the insurgents to extract surrenders if their morale is low. It can help to give indication of a jitter action by the insurgents generally to boost their sagging morale.

Genuine information about the strength of the militants, the size of groups they operate in and the weapons and equipment they use can lay guidelines for the strength of the army required in the areas and will also influence the size of the teams to operate in.

Information about the areas of operation of the insurgents will dictate the areas which require additional stress while gathering specific intelligence and stepping up counter-insurgent operations.

Generic information can give an accurate picture of the attitude of the local people towards the army and the extent to which they are willing to support counter-insurgent operations. This will also dictate the strategy to be evolved for the conduct of psychological operations, its themes and target sections of the population.

The effect of ongoing operations of the security forces against the insurgents gives an interesting insight into fresh initiatives on their part. The image of an insurgent in the eyes of the local people has high value for him. If this suffers, an extra alert for ambushes and counter-attacks is required.

This kind of information also gives details of casualties

suffered by the insurgents in operations. This is a major morale-booster for the army as the level of success of an operation can be judged from this information.

On the other hand specific intelligence or real time intelligence is essential for the conduct of operations. This requires fast reactions, clinical planning, well taped-up battle procedures and battle drills, as the time available for carrying out these operations will be limited. In the absence of specific intelligence it is desirable to try and predict the location of the insurgents by forming a pattern from the generic information and specific but series of delayed information already available. These calculated ventures are expected to pay better dividends rather than mere conduct of operations without the requisite spade work. Feint operations can also be planned and executed to reduce the alternatives and improve the chances of encounters. The process of sifting real time intelligence from generic intelligence requires a systematic approach and careful study of the fluid situation in a regular manner. The intelligence diaries and information collected should be collated and sifted to get the advantage of achieving the probable location for conduct of operations.

In LIC it is imperative that basic combat units like rifle companies and intelligence agencies do not work at cross-purposes. It has rarely been experienced that intelligence gained at various levels and by various agencies is shared by the combatant units. This not only negates the hard work and efforts of the agencies, but it leads to working in water-tight compartments.

There are a number of methods of gaining intelligence at the company level. These methods will vary with the area of operations, level of support to the insurgency, the reputation of the army and the stage of insurgency. However the basic principle of gaining intelligence remains the same. It can be suitably adjusted to the personality of the Company Commander or junior leader in charge.

Listening into the insurgent transmissions is an effective means of gaining intelligence. The intercepts give valuable information about the state of morale, impending operations and casualties of the insurgents. This should however be tallied with information gained from other sources to ensure that false transmissions are not being resorted to.

The utility of reliable interpreters cannot be under-estimated. It is essential that every party moving out has an interpreter and another remains permanently at the post.

There will always be a number of people who would have been harmed by the insurgents in LIC. The feeling of revenge can be buttressed by suitable rewards to gain intelligence about the insurgents. Unless the source is willing to work openly, it is the most important and also the most difficult task of the Company Commander to conceal the identity of the source, but this must be done even at the cost of catches.

There will also be small-time crooks or helpers to the insurgents who run away from the groups. These low-level cadres are extremely useful in tracking the insurgents, giving valuable information about the group leaders' hideouts, caches and publicity material.

Luring the people is another means of getting real time information. Besides money and food it has been seen that jobs are the best means of getting work done. This gives them the security for a lifetime and they do not mind taking risks for it.

One notices that there are a number of groups among the people. These can be based on party politics, religion, land disputes or ideological differences. These disputes can: be used to gain information about rival factions and insurgent groups. However the chances of false information to fix one another is very common and should be guarded against.

One will notice that the basis of intelligence collection rests upon the age old principle of Chanakya. *"Sam, Dam, Dand, Bhed"*, the principle advocated by the wily adviser to Chandragupta in Arthashastra holds good even today. What is needed is the will to exploit it.

The failure of intelligence to gauge the extent of support of the people to the insurgents vis-a-vis the security force has a telling effect as discussed. The Ambela campaign is an example of this aspect.

The operations were undertaken with a view to attacking and dispersing a settlement of Hindustani fanatics at a place called Malka, north of the Peshawur valley (see the Plan). The plan decided upon was to fall upon the place from the north by a sudden march through the Ambela defile in the Chamla valley, so as to drive the enemy towards the Indian plains and into the arms of troops advancing from about Sitana.

The expeditionary force under Sir N. Chamberlain was advancing slowly into the Ambela defile, a long and difficult pass through the hills, when the Buner tribe, a warlike mountain race dwelling in the hill country north of the Chamla valley, suddenly assumed a threatening attitude. For a march from Ambela to Malka with so formidable an enemy on the flank, a far larger army than that told of would have been necessary, and the consequence was that Sir N. Chamberlain found himself obliged to halt in the defile. The Hindustanis who were blocking the exit were supported actively by some small Mohammedan tribes, and were supported passively by the Buners. The inaction of the British force tended to encourage the enemy.

For nearly a month it was in close contact with a resolute and enterprising foe, and severe fighting took place in which considerable loss was suffered. In the end the Buners were conciliated, and the hostile gathering in the defile being dispersed,

THE AMBELA CAMPAIGN

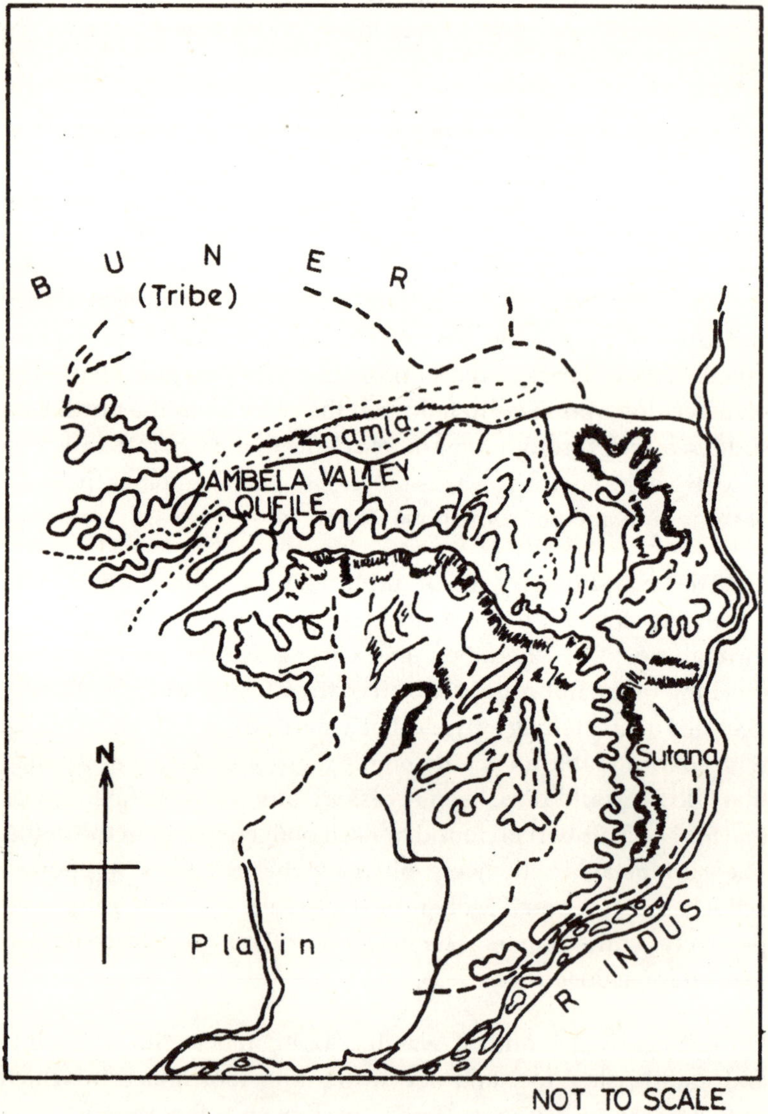

the column marched to Malka and carried out the programme decided upon at the outset.

Another case study gives a graphic first person account of an operation in Doda District of Jammu & Kashmir. This operation gives an example of specific and reliable intelligence as a battle-winning factor in LIC.

Op Rheni

It was 1900 hours and the sky was just turning orange both in the west and east because of the setting sun. I stood alongwith Bisht on the terrace admiring the colours, thanks to the setting sun. The sentry came up and said that Ram Singh wanted to meet us. Getting the "okay ", he went down and up came Ram Singh. He was looking more than pleased with himself and we knew that something was in the offing.

Ram Singh came forward and said, "Saab namaste. Should we sit?" Getting the hint I replied "Certainly, certainly". As we sat, in my usual characteristic opening I said, "So, what is the news?" Ram Singh, finding it difficult to hide his enthusiasm and extending the usually short "saab" said "Sa. a.. b well every thing is fine, but I want you to meet someone." I replied cautiously, "H-m-m, what about, Ram Singh?" He replied "Saab, he may give some useful information". I gave up trying to get him to give me the facts and said "Fine, when should we meet?" Ram Singh as if waiting for the go-ahead replied, "He is presently sitting at my house and I can get him over immediately." Giving him a nod of approval I told him to get the chap to my room.

Mohammed

Just as we were sitting down in our room, getting the chairs arranged, Ram Singh entered alongwith a young man of about 30 years. He was fair, with a light-coloured beard, thin and of medium height. "Saab, this is Mohammed," said Ram Singh. Making a gesture towards the chair I said, "Sit down, so what is the information?"

OP RHENI

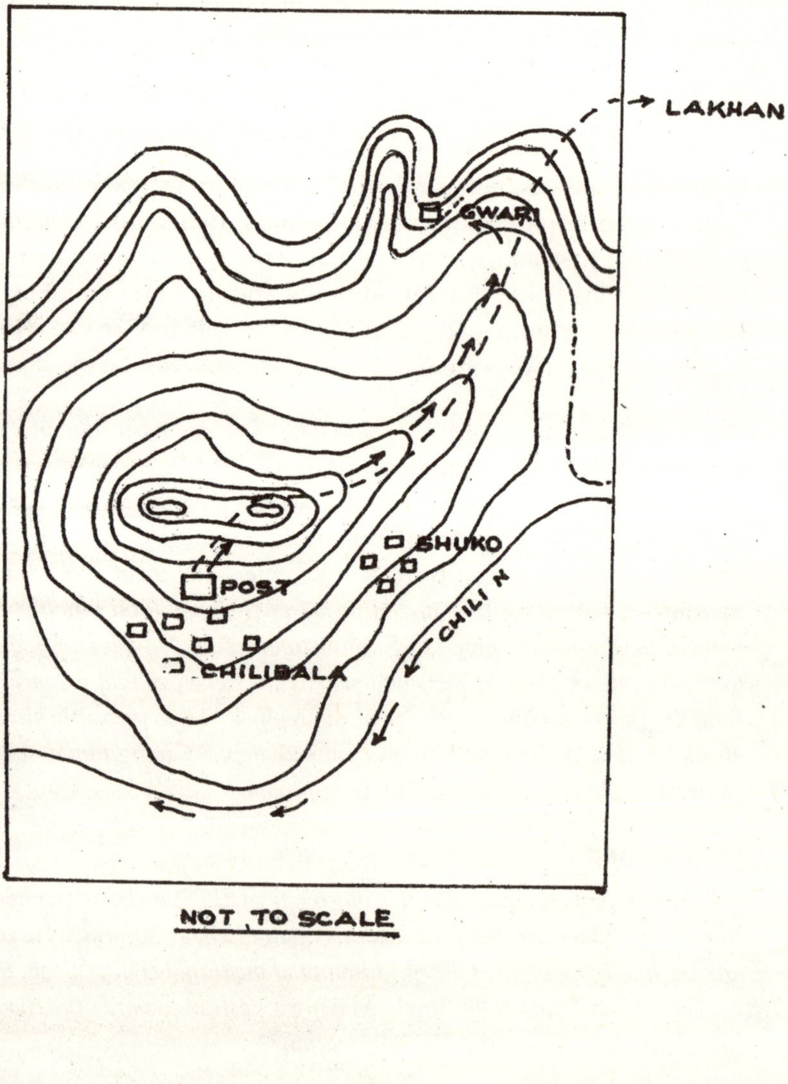

NOT TO SCALE

Mohammed said "Saab, there is a boy from Shuko-Asgar". "I know that", I said.

Mohammed continued, "He is presently in his gwari alongwith his mother. I had seen him the day there was firing at Chilipai coming up the track towards Shuko with an SLR and two grenades. Recently I had gone to his gwari in Rheni Forest and again I saw the weapon with him."

My interest increased and I asked him, "But how long back have you seen the weapon with him? And how can you be sure that he will be sitting there with the weapon? The militants must have taken it by now."

He gestured impatiently with his hands and said, "No, no, I have seen it even after that and tomorrow morning again I will go and confirm that Asgar is there. I have information that he has been injured in the previous encounter."

I was satisfied with the news, though I could not digest the fact, and Bisht agreed with me, that a weapon could be with an injured insurgent for so long and that too an SLR. Since we had no choice, Mohammed was asked to confirm the news and come back the next day.

The very next day after confirming the information Mohammed again came at approximately 1700 hours. He was absolutely sure of himself and his enthusiasm got the better of us. We gave him the time out as 2000 hours. He moved out promising to meet us at the assigned location.

The day never seemed to end. I and Bisht paced the room anxiously waiting for the sun to set and the operation to commence. Capt Bisht and Subedar Patil were detailed for the operation. It was a pitch dark night, and as is the drill, we did not put on the torch at night. Though the move was not over a very long distance, the difficult jungle terrain and the slushy route made it extremely dangerous. As Bisht relates the incident today, there was no one in the party who did not fall at least once. But not being deterred by the successive and painful falls, the party moved on to its destination slowly but carefully. Night moves are very peculiar. Firstly, a distance which can be covered in half an hour in the plains may take two or even three hours at night. Time, just seems to freeze and there is a, feeling of nothingness. Secondly, as one cannot see anything, each step is a risk. The, fear of hurtling down a steep cliff is ever-present. It is tension and fear of the unknown that grips the mind. But at the same time, responsibility and a sense of purpose drive the mind and the body overcoming all odds. It is a combination of training and mental tenacity which keeps driving the body.

I am sure that the condition of the party was similar. They were also aware that since the gwari was in a totally inaccessible area the chances of the presence of militants were quite high.

With due care and caution the party made its way through the Khorna Gali towards their destination on the track going towards Lakhan. After a climb of approximately three hours, the main track was left, and approaching the gwari the party started moving down a steep and slushy slope. Despite the numerous falls, the steadfast approach took the party to the close vicinity of the gwari. As the party neared the gwari suddenly the militant's sentry opened fire. The chatter of AK-47 fire in the still of the night announced the presence of militants in the gwari.

Training, steel-strong guts, mental robustness, physical courage and the spirit of a proud, brave Company came to the fore. Captain Bisht, taking stock of the situation and making a quick tactical evaluation, divided the party into three sub-groups. One under Subedar Laxman Patil, the second under Sepoy Bhishumbar Singh and the third under himself.

Quick and concise orders were given to the party commanders and in turn they quickly briefed their parties as per the plan. Firing continued from both the sides.

I think a number of times as to what is it that drives a soldier through the hail of bullets onto the objective. With the dark, enveloping clouds of death fast approaching the vulnerable soldier, why do the thoughts of his family, his home, his loved ones not stop his feet and weaken his resolve. In a world of selfish and materialistic people, why does a soldier brave the dangers of death for a paltry sum?

The firing continued for about half an hour and thereafter the militants fled from the scene of action. The three parties closed in and broke into the house. Asgar was captured after a brief scuffle alongwith an SLR. Chase was given to the militants but they managed to get away.

After the area in the close vicinity of the gwari had been secured, spot-interrogation of Asgar was carried out. Thereafter the party started moving back to the camp.

All the tiredness and fatigue had vanished and it had given way to exuberance and happiness. The first success is always the sweetest. The party was quicker on the move now and was eagerly waiting to reach the post and share the happy news with the rest of the Company.

Meanwhile, in the Company, I, and I am sure the rest of the personnel, had not spent a more tension-filled night. At the slightest noise I would get up and go to the entrance to find Ajay, my radio operator and Havildar Sheikh Hussein Sab already there looking out into the nothingness waiting for positive news from our party. For the first time I realised as to why parents are worried when children get late while coming home. I could feel the pressure playing on me. I have never experienced such pressures even during the most dangerous of patrols or operations.

At last the wait was over as Ajay came and said, "Saab, call from Bisht Saab." I sprang out of my bed and went out into the balcony snatching the handset and called,

Alpha/Alpha "1 for 2, Tiger listening, over." I heard Bisht.

Alpha/Alpha "2 for 1, we have got Asgar alongwith one SLR and magazine, over."

Alpha/Alpha "1 Roger. Well done, hope everyone is fine, over."

Alpha/Alpha "2 yes, everyone is OK. We will be with you in fifteen minute, over."

Alpha/Alpha "1 roger. Where are you speaking from? over."

Alpha/Alpha "2 Khorna, over."

Alpha/Alpha "1 roger. Waiting for you, out." The wait was soon over as I saw the scouts approaching the camp. Bisht came in next. As he came to the first floor, I hugged him and congratulated him for the success. He again went down to take the report. I shouted to the CHM (Company Havildar Major) to get tea ready for the party and went and sat down in the balcony.

I felt as if a big load had been taken off my shoulders. All of a sudden the cool early morning breeze became sweeter as the balmy air rustled my dishevelled hair. The twinkling sky looked like a diamond-studded necklace, and the moon which was just about appearing, looked like having joined in our celebrations. Bisht having broken off the party with the traditional break-off war cry of "Shankar Bhagwan ki jai" came over and joined me in the balcony. The rising moon was enough to see the beaming face of Bisht, who was satisfied with the success.

Today, immediately after our arrival at Chilibala we had been successful in our sworn endeavour.

Note

1. Frank Kitson, *Low Intensity Operations.*

Communications

LICs guarantee a fluid operational scenario where flexibility is very essential. In this environment, a reliable communication network is necessary to convert the vision of a commander into reality through the sub-units. A communication failure cannot only cause operational failure, but it can also result in the loss of lives for want of support in time.

Besides the quality of equipment in use, which faces the brunt of severe criticism, it is the ingenuity in its usage which is more relevant and important. There have been instances where the ANPRC-25 has communicated over 200 kms with an improvised wire type of antenna and the VA set over 40 kms.

Insurgent ambushes and roadblocks will require quick reinforcements. This can be done by establishing communication with either the nearest unit or the HQ. It is sensible to keep frequencies of all the units in the area including the paramilitary forces. During emergencies their QRTs and relay facilities can prove to be very useful.

Wherever possible communication should always be duplicated. Though this is an age-old principle, yet case studies often suggest laxity in this regard distribution of radio sets in a party also assumes sumes significance. There have been instances where maximum radio sets were found in a particular vehicle or platoon resulting in the others getting cut off. This severely impedes the reaction capability of the sub-unit and the eyes and ears of the Company Commanders are non-functional due to the lack of feedback.

The lack of the correct use of RT (Radio Telephone) Procedure and secrecy codes often lead to security loss and pre-emptive strikes by the insurgents. This laxity comes in because of prolonged stay in LIC and extensive use of the radio sets. Soon the radio set replaces a telephone, being the only means of communication. This habit has frequently been noticed with radio operators, who find the radios as the only means of entertainment in the secluded atmosphere. These habits should strictly be checked and discouraged.

In mountains, crests gain special significance. These heights serve as the relay centres for the communication network of the insurgents and an impediment for the transmissions made by security forces from posts at lower heights near road-heads. In some cases where feasible, relay centres have been established to overcome these problems at least during operations and LRPs. While planning routes these crests should be borne in mind to ensure that there is no communication failure during the course of operations.

With the advent of Radio Relay, communications have received a major boost. Wherever these facilities are available, secure and efficient communication helps achieve greater efficiency. During operations this should be made use of and wherever it is not available bids can be placed for its installation.

Telephones give an added benefit and means of communication. These are ideal means of communication for routine conversations. However wherever the cable passes through hostile areas, the dangers of tapping are present. Operational conversation should be avoided in all such cases.

A connected problem of telephones based on WD cable (a

decades-old type) is the frequent move of line parties for the maintenance of the cable routes. These are specially vulnerable to ambushes and sniper fire from the insurgents. The security of these parties should be ensured. OP PAWAN has a history of a series of ambushes of these line parties.

Civil telephones also serve a useful purpose in giving an additional dimension to communication. Though these are not secure, they can be used to communicate administrative requirements. Besides this, telephones are very beneficial for posts established in remote areas at extreme distances from the HQ. These telephones give access to the Tehsil officials and police which can be of use for psychological warfare operations and civic action projects.

The results of failure of communication have numerous case studies as proof. An example from a campaign of 1897 in mountains in the 'Tirah' sector is a suitable incident to prove the relevance of the principle of communications in LIC.

Mistakes as regards orders have frequently caused difficulty and loss in hill warfare, owing to isolated parties misunderstanding the intentions of the Commander and marring his plans. And before suggesting how best to avoid such situations an example from the campaigns of 1897 on the Indian frontier may be given, showing what unfortunate results may occur due to the non-receipt or misconception of an order.

Practically the last serious combat of the Tirah campaign ended in a disaster owing to this (communication). A column had crossed the Shin Kamar Kotal, and some of the troops had advanced a little way down the valley on the far northern side. The route on either side of the kotal was overshadowed by hills on the flanks, the whole depression running north and south. The enemy showed in considerable strength on the hill tops, but there was every reason to believe that the return journey would have been attended by not very serious fighting, had not a company, owing to some mistake which has never been quite explained, withdrawn from a knoll and been promptly seized by the enemy, and the retirement of the troops beyond was so seriously jeopardized by its loss that it became necessary to recapture it at all costs from strong hostile bodies elated by an apparent success. The knoll was stormed, but there were many casualties, and the troops forming the rear guard thus found themselves encumbered with numbers of wounded In the final withdrawal from the knoll, the troops, hampered as they were by the wounded suffered further serious losses, and the rear guard was in the end only extricated with the utmost difficulty from its awkward situation. In this unfortunate affair, 5 officers and 22 men were killed and many were wounded, among the officers killed being Lt Col Haughton.

The first attempt to cow the Mamunds has been referred to in an earlier paragraph. The day's fighting is remarkable for the number of lessons in the art of mountain warfare which it teaches, and one of its most memorable incidents was the unfortunate result which attended the non-delivery of an order to a certain company at the proper time. After having been heavily engaged ,for some hours the brigade retired, keeping the enemy at a respectful distance at most points. But a Sikh company on the extreme right had pushed high up into the hills, and before the order to withdraw eventually reached it, had become completely isolated and was being hard pressed by very superior numbers. To extricate it the guides had to turn back, and it only reached the foot of the hills towards dusk with both officers wounded and short of ammunition, having had a desperate scuffle. It was largely due to this contretemps that the whole brigade was benighted, and that, while most of it regained camp in detachments, some troops were cut off and remained in a situation of much danger till past midnight.[1]

It is obviously important to devise, if possible, some method of avoiding such unfortunate occurrences; but this is not easy. The ordinary precautions tending to ensure the correct transmission of messages should of course be observed — messages should be written down, should if practicable be sent twice by distinct orderlies, should be clear and concise; if the receiver of a message is in any doubt as to its meaning he should if possible refer back to the sender, and so on.

Note

1. Callwell, Col C E, *Small Wars,* pp. 311-22.

Axis Of Maintenance

In protracted wars like LIC, where permanent posts are established to form a counter-insurgency grid in an area, it becomes imperative to maintain an effective logistic back-up management system, despite the constraints of weather, roads or lack of the same and threat from the insurgents.

At times it is not realized as to how important a role sound logistics back-up plays for the morale of the troops. The realization that someone is risking his life to feed the Company can do wonders for the army. The quality of rations received by the army during OP PA WAN did a good deal for our morale and gave us the motivation to perform better. Lt Gen S C Sardeshpande says:

"It was in logistics that I think the IPKF performed admirably. It was untrodden ground without a precedent to help us we had no experience of supporting logistically a large composite force of all arms across the seas in foreign country, which not only was hostile but also lean in resources. We had to lug everything across the sea by ship or aircraft, often both we repaired the airfields, constructed temporary structures for housing huge stocks and unimproved ports where our ships berthed and picked up, or discharged cargo. All the three services-the Army, the Navy and the Air Force-representatives and staff would sit down every day and plan the transit and transport of men and material. I am sure that had it not been for this efficient system of logistics back-up, the IPKF would have been forced to pull out much earlier."[1]

In areas of the North-East and Doda District of Jammu & Kashmir, posts are extended over long distances and are often located at places where there is no provision for movement by roads. It is here that the efficiency of the logistics back-up is tested. Monsoons in the North-East and snow in Doda make the situation worse. It forces the post to carry out monsoon, winter stocking which itself is an entire exercise. During this period it is imperative to ensure the safety of the frequent patrols which carry maintenance stocks. While carrying out stocking the following points need emphasis:

- The timings and routine of the patrols should be varied.
- A strong protection party should accompany the administrative convoys.
- The parties should he briefed every day to ensure that complacency does not set in and security is not compromised.
- It should be ensured that the party is in effective communication throughout the duration of the patrol. Often these administrative convoys are targets of insurgent ambushes.
- Road opening should be carried out irrespective of the vast distances and periodicity of move.
- The move of the party can be so tied up that it moves just behind buses or trucks to avoid the threat of IEDs.

These points gain added emphasis during the stocking period, but even during the rest of the year these basic precautions should be carried out. The Indian soldier has the capability to withstand extreme deprivations, hardships, difficulties and stress. It is a rare quality in this materialistic and selfish world today. It is this quality which helped in crushing the Japanese offensive during the Second World War and it is this very spirit which is controlling areas of insurgency against all forms of attacks, psychological and physical, despite restrictions. We as Company Commanders should not aim at testing this spirit further, we must ensure that without the limitations of the area our command is provided the best possible logistic support. It is not their privilege but their right.

The Nile Expedition was a colossal effort at maintaining the supply line. The campaign was conducted more than a century ago, but the relevance of the principle of a safe and efficient axis of maintenance has not diminished.

Like many other small wars, it was essentially a campaign against nature. A struggle against the difficulties arising from supplying the wants of troops traversing great stretches of desert country. But it had this peculiarity-it was a campaign not only against nature but also against time, because it was undertaken to relieve a beleaguered city. For this reason, and also because of the remarkable manner in which its phases demonstrate the influence which the question of food and water may exert over a small war, an outline account of it from the point of view of supply will not be out of place.

From Lower Egypt to the second cataract at Wadi Haifa, the railways along its course and the numerous steamers upon its waters make the Nile and its valley an excellent line of operations for a small army. For this reason the Nile Expedition may be taken as having had its base at Wadi Haifa.

Above Wadi Haifa up to Third Cataract (see the plan) the Nile, at the season when the despatch of the force really commenced, and for several months afterwards, forms a series of rapids most difficult to navigate. From the Third Cataract to Hamdab the river offers no difficulty to small steamers; but only two were available on this stretch. Armed opposition was not to be expected north of Korti, therefore between Wadi Halfa and that point the transport of the troops and their food, and the arrangement of supplies for feeding them en route, were practically the only points to be thought of. The campaign proper would only begin after passing Korti.

Lord Wolseley's plan was to start his force from Korti with 100 days' supplies, which, supplemented by supplies which would be conveyed across the desert from Korosko to Abu Hamed as soon as that point had been reached, it was hoped would carry the army from Korti to Khartum. The force was to row itself from Wadi Haifa to Korti in the boats specially sent out for the purpose, carrying its 100 days' supplies with it and fed by the way from depots established in advance along the route. A small mounted detachment and the specially organized camel corps, however, marched the whole way.

The first thing, therefore, was to push on supplies by every available means so that the whole force should be fed on its way to Korti, and so that there should be sufficient supplies at the front for the mounted troops (who had not the boat supplies to depend upon) to be able to carry out such operations as might prove necessary. In the meantime most of the troops were held back below Wadi Haifa up to the last possible moment, although a small force had to be pushed on to protect the supply depots and secure the line to Korti.

In spite of the most strenuous exertions, the difficulty of collecting the necessary supplies along the route and at Korti gave rise to delay. The force took longer to assemble at its advanced base than had been anticipated, because food for its consumption took longer to collect than had been expected. Time was throughout of the most momentous importance. And so it came about that when the troops eventually began to reach Korti, the situation had become so critical at Khartum that Lord Wolseley decided to at once send the camel corps across the desert from Korti to Metemma, to open up communications with General Gordon. The camel corps had been organized with the possibility of this contingency in view, although the original plan of campaign had been that the whole, force should move by river. The difficulty of calculating the time within which supplies can be stored along a line of communications, even when the enemy has not to be taken into account, was thus demonstrated before the actual campaign began.

The force detailed to march to Metemma reached Gubat close to that point and 180 miles from Korti; but it took 21 days to do so. The delay was entirely due to supply difficulties, for instead of marching straight across, it was found necessary to form a depot at Jakdul about halfway where there were good wells, and to send the transport back from there to Korti to bring up more food, before the rest of the distance could be covered. The fight at Abu Klea had a most unfortunate effect as regards supply, for the transport suffered considerably in the action; and the transport being throughout heavily worked and getting little rest was seriously crippled by the time that the, force had reached its destination near Metemma.

Khartum having fallen immediately after this and a complete change having thus occurred in the strategical situation, that desert force after a time withdrew to Abu Klea. Lord Wolseley had for a time hoped before closing operations for the season to capture Berber by a combined movement of the desert force and of the River Column which was in the meantime following the Nile, and thus to establish an advanced post at an important strategical point in anticipation of a campaign some months later. But it was found that the transport of the desert force had completely collapsed under the strain thrown upon it, and that the necessary food for the force intended for this enterprise could not have been carried. Owing to the supply question the desert force had thus become inoperative.

But the River Column had also become practically inoperative, likewise on account of supply. This advanced by boat with its 100 days' ration as originally intended, but, owing to an exceptionally low Nile, the rapids to be ascended proved to be more difficult than had been anticipated, and the progress of the column was unexpectedly slow. A convoy had been organized to move from Korosko to Abu Hamed and replenish supplies; but before the column reached that point it become clear that, even allowing for this assistance, the supplies would not suffice to carry it to Berber and back-the already referred to loss of a month's biscuits very seriously affecting the question. So it became necessary to. recall the column to Korti, and the operations came to an end.

It may seem strange that the conveyance of a large store of food from Korosko to Abu Hamed should have offered no serious difficulties,

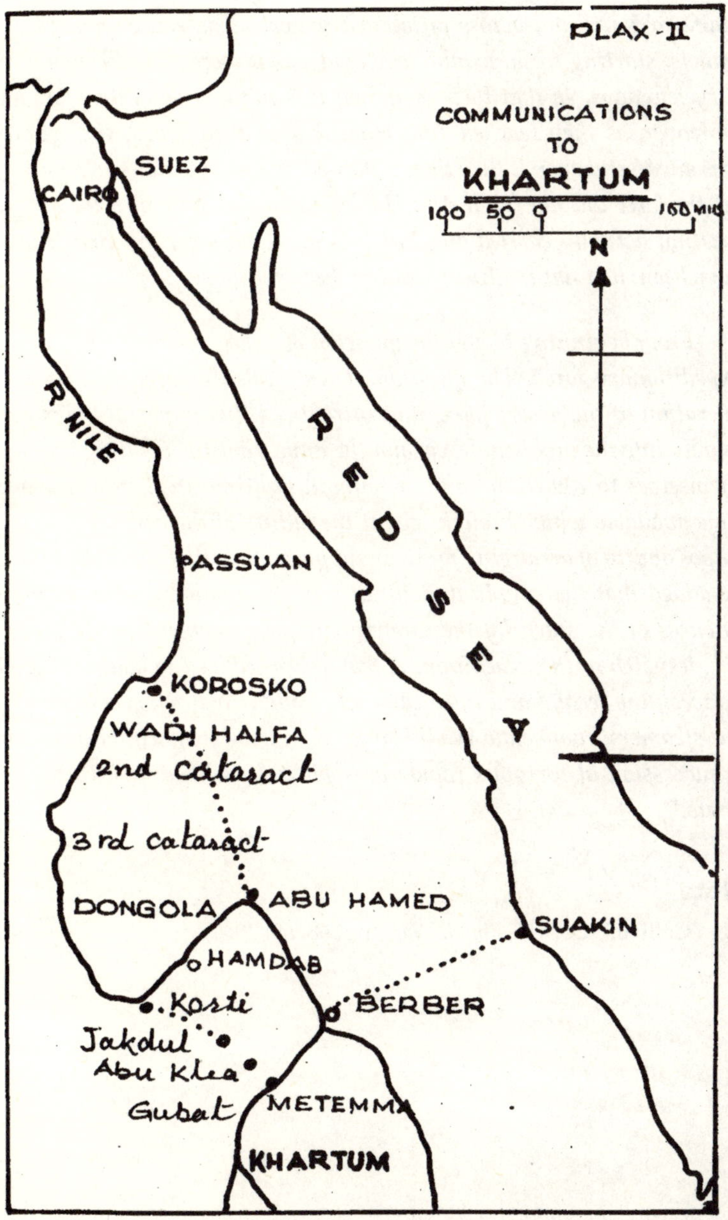
PLAX-II
COMMUNICATIONS
TO
KHARTUM
100 50 0 150 MIL
N
SUEZ
CAIRO
R NILE
RED SEA
ASSUAN
KOROSKO
WADI HALFA
2nd Cataract
3rd cataract
DONGOLA
ABU HAMED
HAMDAB
SUAKIN
Kosti
BERBER
Jakdul
Abu Klea
Gubat
METEMMA
KHARTUM

considering the great distance, 230 miles-two-thirds of that from Wadi Haifa to Korti and across an almost waterless desert. But in case the camels, starting fresh, would have made, as it were, a forced march in a very few days, so that their own food was not a serious item. As long, moreover, as they reached Abu Hamed with their loads, the operation was successful even it they then collapsed. Such conditions seldom occur. As the force did not reach Abu Hamed this notable item in the program was not actually carried out; but it is practically certain that it would have been, had not the River Column been recalled.

From beginning to end the question of the movements of the expeditionary force. The question of water also greatly influenced the operation of the desert force, and introduced this important element of supply info the problem. It was not the impossibility of getting the troops themselves to Khartum, with or without fighting, in the short apace of lime available which brought about the failure of the campaign, but the impossibility of arranging for their supply within that time. And it must be added that the supply difficulties were enormously increased by the lateness of the start, by the unfortunate postponement on deciding on the despatch of the expedition. A few weeks sufficed to convert the Nile between the second and third cataracts from a great waterway up which the steamers from below Wadi Hal fa could have steamed with ease, into a succession of tortuous rapids passable only with difficulty by small boats.[1]

Note

1. Callwell, Col C E, *Small Wars,* pp. 68, 69, 70.

Ground

Confirmation of the ground is of great assistance in battle. Therefore to estimate the enemy's attempts and to calculate the degree of difficulty and distances of terrain in order is to control forces of victory task of a superior general. He who fights with full knowledge of these factors is certain to win, and he who does not is sure to be defeated.

— Sun Tzu

Any area where the army operates for CI operations belongs to the insurgents. They have intimate knowledge of the ground: every nook, corner, fold, defile and cave is known to them. The advantage of ground positively rests with the insurgent. In these circumstances initially the army finds itself totally incapacitated by the sudden disappearances of the insurgents during a chase. Numerous instances of the army missing an insurgent in hiding at just a few feet have come to light. This situation is not only peculiar to the jungles of Manipur, Nagaland and Assam, but also to the built-up areas of the valley where the ingenuity of the insurgents have given a tough time to the army.

This principle of LIC continues to play havoc with the army's serious and concerted effort put in to overcome this disadvantage. There is no short cut to familiarization with the area. It is a painstaking continuous process which requires extensive movement in the area of operations. It not only involves knowing the roads, lanes and by lanes, but also the jungles and swamps.

The principles of Manoeuvre and Flexibility are based on the ability of move swiftly, through the shortest route by day as well as by night. All other advantages in terms of training, firepower, strength get neutralized when a Company Commander does not have the ability to place the troops at the desired place within the given time-frame. The knowledge of ground will also lead to areas which are suitable sites for ambushes by, and for the insurgents.

It is essential to know the places to spring ambushes from and to be careful while moving through the area. In mountains special care must be taken on constricted routes with steep or rocky slopes on the sides: though these may not be ideal locations to fire from, yet they are extremely dangerous because they afford no chances of counter ambush drills.

During the initial period in an area, it is advisable to take local guides or sources along for guidance of the route. These guides should be reliable and fit enough to keep up with the tough routines of the company.

Sun Tzu gives an interesting analysis of ground and classifies it into six varieties. He calls the first kind 'accessible' which both the insurgent and the army can traverse with equal ease. In this kind of ground whoever takes the sunny positions and keeps his logistic route unimpeded has the advantage. The second variety is 'entangling' which is defined as easy to reach but difficult to get out from. In this type if the insurgent is unprepared and the army 'sallies' out, the army wins. Conversely if the insurgent is prepared and the army sallies out, but does not win, it is not profitable. The third variety is which is equally difficult for both the insurgent and the army. This is called 'temporizing': this nature of ground is such although the insurgents hold bait, the army does not take it, and marches off to entice him. And when half his force has been pulled, the army strikes from an advantageous position. If the army occupies narrow passes, it must wait for the insurgents. If the insurgents occupy these, they should not be attacked unless they have not occupied them fully. On precipitous ground the army should take the sunny heights and await the insurgents. If the insurgents occupy it the army does not attack and marches off. When at a distance from the insurgent and with equal strength, it is difficult to provoke battle, and unprofitable to engage the insurgents.

These ideas about the use of ground have been expressed approximately 2,400 years ago. Yet their relevance is strikingly evident even today. They give insight into the planning of ambushes, opening of fire at long ranges, value of an advantageous position, dangers and advantages of terrain features like defiles, steep slopes and difficult slushy areas.

Moral Ascendancy

It is evident that in our society public opinion is a notable source of power. If leaders have the support of public opinion, their freedom of action is enhanced and obstacles become surmountable. When support of public opinion is sharply withdrawn, public figures topple, laws become unenforceable and bastions of economic power tremble.

— John W Gardner

On induction into the area of operations, the army usually finds itself unwanted. The resentment of the people is only too obvious. It is against this backdrop of hatred that the army has to establish itself, to carry out its task. In LIC where fish cannot survive without water, it is the support from the water vis-a-vis drought that both the constants have to fight for: the insurgents and the army. Whoever gets water gets transformed into the victor, and who faces drought gets vanquished.

It is not so much of operational prowess, and military might, that wins the people. It is the humane attitude and moral ascendancy which transforms the hostility of the people to cooperation. Though the army is known for its positive attitude, yet rare incidents of high-handedness can destroyer his image and prove counter-productive, in consonance with the plans of the insurgents.

The insurgents also realize this fact and go in for popular measures and schemes to win over the people. Orders against illegal production of liquor, smoking, exorbitant expenditure in marriages, bribery, etc. have been enforced both in Punjab and Kashmir. The guidelines given to the Vietnamese army soldiers by Ho Chi Minh aimed at winning over the people. Surprisingly these principles are not only applicable but will prove to be as effective in our scenario too.

Twelve Recommendations

— *Six Forbiddances*

1. Not to do what is likely to damage the land and crops or spoil the houses and belongings of the people.
2. Not to insist on buying or borrowing what the people are not willing to sell or lend.
3. Not to bring hens into mountainous people's houses.
4. Never break our word.
5. Not to give offence to people's faith and customs (such as to lie down before the altar, to raise feet over the hearth, to play music in the house, etc.)
6. Not to do or speak what is likely to make people believe that we hold them in contempt.

—Six Permissibles

1. To help the people in their daily work (harvesting, fetching firewood, carrying wood, sewing, etc.)
2. Wherever possible to buy commodities for those who live far from markets (knife, salt, needle, thread, pin, paper, etc.)
3. In spare time to tell amusing, simple, and short stories useful to the Resistance, but not to betray secrets.
4. To teach the people the national script and elementary hygiene.
5. To study the customs of each region so as to be acquainted with them in order to create an atmosphere of sympathy, then gradually to explain to the people to abate these superstitions.
6. *To show to the people that you are correct, diligent and disciplined.*

The order' by Ho Chi Minh terminated in the form of a poem which characteristically gave the importance of his recommendations.

The above-mentioned twelve recommendations
Are feasible to all
He who loves his country
will never forget them
when the people have a habit
All are like one man,
with good army men and good people,
Everything will be crowned with success.
Only when the root is firm, can the free live long,
And victory is built with the people as foundations.

Similarly the aspect of moral superiority is emphasized at the highest levels in the army. The commandments issued to all personnel of the armed forces in insurgency is a constant reminder in the do's and don'ts for a soldier

A Company Commander must realize that the activities of the Company not only project the Company but they project the image of the Army and the Government. This should be repeated time and again to the personnel. With increased contact with the corrupt system, a soldier at times is induced to threaten his way out of paying for stores purchased, or pick up a valuable during searches. These malpractices are at times taken as privileges of fighting for the people. This misconception should be removed from their mind, and strict disciplinary action should be initiated against defaulters. There may also be cases of misbehaviour with women, though it is rare. Such individuals should be given exemplary punishment. Besides, the aim should be to avoid such situations by minimizing the contact of the Company with the local people.

It is suggested that a committee should be formed under the senior JCO which monitors all monetary dealings and purchases made by the company personnel. At the end of the month, a report should be taken to ensure that all payments have been made.

All these methods, and many more, aim at giving a contrast between the looting of the insurgents and the monetary help of the army, the misbehaviour of the insurgents and the girls tying 'rakhee' to the army personnel, the killings of the insurgents and the medical help of the army. It is these contrasts that will prove the noble intentions of your Company. This principle of LIC may not directly help militarily, but it will make a stupendous moral impact, which will eventually win the people, and the insurgency.

Media Management

Terrorism relies heavily on propaganda by deeds and declarations, in attempts to demoralise the opponents and sap their will to fight, and in the process creates a psychosis of fear.

—Jagmohan

Media Management as a principle of LIC despite its tremendous ramifications is the least used principle by the army. It also happens to be the most often used (rather misused) weapon of the insurgents and their supporters like Pakistan to let loose vicious and false propaganda. Though this medium is known to have immense possibilities and capabilities, its use is yet to fall under a well-coordinated and directional approach which can counter the propaganda of the insurgents.

Newspapers, magazines, posters, pamphlets, leaflets, radio, television, word of mouth-all form the multidimensional arms of this octopus. Its mind-boggling exposure can be estimated from the kidnapping and killing of eleven Israeli athletes during the Olympics in September 1972 by the terrorist group Black September which was watched by an estimated 500 million television viewers all over the world.

Jammu & Kashmir is an example of the use of this principle by the insurgents. Jagmohan's diary brings out this aspect and he says.

As the shadows of terror ism lengthen over the Valley, the vernacular press, with negligible exception, became its willing instrument. A part of it was afraid of the terrorist 'bullets, a part was not otherwise disinclined to play the subversive role, and a part merely whipped up its foundational motivation of propagating the Islamic Fundamentalism ideology. The end result of all this was that the entire vernacular press became a mouthpiece of the subversive elements. It virtually abetted, incited rationalised and even glorified terrorism.[1]

Are we prepared to answer this degree of misuse of the media Are we not handing over this principle of LIC to the insurgention a plate to use it against us? Perhaps we are.

The answer to the problem lies in considering media as an important aspect of LIC and give it the status of an organizational aspect in its entirety managed as a full-time job with the best available brains in the army and outside. If we all feel that insurgency is a national issue which threatens the unity and integrity of the country, I do not see why the very best in civvy street will not want to come and give a helping hand.

It is desirable that specialist training should be imparted to army officers in reputed organizations and institutes where mass communications can be taught as a subject. Interaction with full-time journalists and news managers will help understand the colossal power of this medium. The Times Group runs one such course which can be useful for the army.

At present an organization does not exist at each level of command to manage the reporting system. Either reporters, or qualified officers, should be available at each level down to the infantry battalion to work out this aspect in consultation with the superior headquarters.

The effects of serials like 'Fauji' and 'Param Vir Chakra' are common knowledge. Similar image-building exercises will help create national awareness about the difficult work conditions of the army and create mass support for its activities.

We are often accused of being too secretive about our lives and in the activities. Perhaps the time is ripe to open up at least LIC scenario. The fourth dimension of the modern battlefield awaits the use of its unlimited resources. Are we ready to use it?

Note

1. Jagmohan, *My Frozen Turbulences in Kashmir,* pp. 387-88.

Part IV
Tomorrow

Tomorrow

What does the changing complexion of the new world order have in store for us, the armed forces and in general the country? It is almost 25 years since the time India fought a conventional war. It is half a century ago that a world war was fought. Despite and inspite of these obvious realities, the role of the armed forces has only increased, ironically!

Seeing statistics and daring to predict based on the revealing environment, the chances of a world war are remote. Similarly the chances of a conventional war in the Indian subcontinent must certainly be one of the areas marked in 'red' for the pundits of future conventional war predictions. If this be the scenario, where 'High' and 'Medium' Intensity wars are fast taking a back seat, surely the 'Low' Intensity variety promises to have a 'bright and vibrant' future in the years to come. The changing face of socioeconomic transformation is causing an upheaval within the very borders of nations, threatening to change the nationalistic boundaries that govern them. Martin van Cleveld says, *"As war between states exits through one side of history's revolving door, low intensity conflict among different organizations will enter through the other."*[1] Having safely concluded on the relevance and increasing importance of LIC in the future, what role can the armed forces play in this scenario? The answer hinges on the vital issue of prioritizing the role for the armed forces. On one hand a nuclear state provides the necessary deterrent to safety and uses the expertise of the armed forces for LIC; and on the other hand a non-nuclear state is forced to maintain the forces in a fairly high state of alert against possible aggression.

Does the future have a semi-policing role in store for the army where, devoid of the unconventional war threat, LIC becomes the top priority? Is the Rashtriya Rifles model set to engulf the army? These and many more related questions will decide the future of the armed forces.

For a Company Commander, the role of the army in LIC can be taken for granted. The situation of insurgency in the North-East and Jammu & Kashmir promises an involvement which cannot be taken lightly. It is high time that we situp and recognise this reality.

On the conceptual and tactical front it has been my endeavour to add my contribution to enable us standardize the LIC battle for a Company Commander; the pivot of military actions. The formulation of the proposed principles of LIC will no doubt require more deliberation and thought before they can ultimately be finalized. Similarly the planning, preparation and study of LIC will require much more effort to make a company commander, ready for the conflict as and when the time comes.

The role of the media and the importance of legitimacy of operations have also been highlighted. In the future, where media has come up as the fourth dimension in LIC, a concerted effort will be required to institutionalise this pent-up force. Its exploitation will undoubtedly play a major role in tomorrow's LIC scenario. Similarly with the increased emphasis of legitimacy of operations, the aspect of human rights will merit more deliberation. The salient aspects will require careful study and subsequent dissemination to the lowest levels of the command structure.

The Army has not gone in for any major modification in its organization or equipment authorization in the LIC environment. This may become necessary because of the peculiar role played by the Army in the years to come. For a Company Commander, the requirements in tomorrow's LIC scenario will be:

- Helicopters should be made available for operations and logistic support. The minimum level at which these flights of helicopters should be available is the brigade. The recommended scale is a flight. Besides this helicopter gunships should also be available on call.
- The scale of light vehicles should be increased. Every company should be authorised four jeeps, four one tonners and one three ton vehicle. Besides, a water truck should also be available. This will enable both independent moves of platoons and the company.
- Shotguns should be authorised at the scale of two per section. These can be given as sector stores. The variety should be magazine-fed, with the facility to fire multiple ammunition. An example is the Benelli M 3 Super 90 Convertible Shotgun manufactured by Italy.
- Secrecy devices for radio sets should be authorised at the company level for inter-company communication.
- Three sniper rifles should be authorised to a rifle company.
- Global Positioning System (GPS) at the scale of 04 per rifle company will be able to give the precise location with a geographical accuracy of 100 metres.
- A doctor (even if he is a civilian) should be posted with each company keeping in view the distances of the companies from the Battalion HQ.
- An intelligence officer should be posted with each battalion to cater for establishing and supervising the intelligence network at the battalion level.
- A media officer should be posted with each battalion and down to the company, if the distances cannot be covered by one detachment.
- Officers and their detachments should be posted to the battalion from Engineers to cater for the field engineering aspects.

— Each Rifle Company should have an ICV grouped with it wherever the nature of ground permits its use.

Note

1. Cleveld, Martin Van, *On Future War,* p. 224.

Annexures

Annexure I

Instrument of Accession of October 26, 1947

With the conditions obtaining at present in my State and the great emergency of the situation as it exists, I have no option but to ask for help from the Indian Dominion. Naturally they cannot send the help asked for by me without my State ceding to the Dominion of India. I have accordingly decided to do so and I attach the Instrument of Accession for acceptance by your Government.

Reply by Lord Mountbatten on October 27, 1947

In the special circumstances mentioned by Your Highness, my Government have decided to accept the accession of Kashmir State to the Dominion of India. It is my Government's wish that, as soon as law and order have been restored in Kashmir and her soil cleared of the invaders, the question of the State's accession should be settled by a reference to the people. Meanwhile, in response to Your Highness' appeal for military aid, action has been taken today to send troops of the Indian Army to Kashmir to help your own forces to defend your territory and to protect the lives, property and honour of your people.

Annexure II

The Resolution of the United Nations Security Council of August 13, 1948

Resolution of the Commission of August 13, 1948:

The United Nations Commission for India and Pakistan.

Having given careful consideration to the points of view expressed by the representatives of India and Pakistan regarding the situation in the State of Jammu and Kashmir; and

Being of the opinion that the prompt cessation of hostilities and the correction of conditions the continuance of which is likely to endanger international peace and security are essential to implementation of its endeavours to assist the Governments of India and Pakistan in effecting a final settlement of the situation:

Resolves to submit simultaneously to the Governments of India and Pakistan the following proposal:

Part 1: Cease-fire Order

A. The Governments of India and Pakistan agree that their respective High Commands will issue separately and simultaneously a cease-fire order to apply to all forces under their control in the State of Jammu and Kashmir as of the earliest practicable date or dates to be mutually agreed upon within four days after these proposals have been accepted by both Governments.

B. The High Commands of the Indian and Pakistani forces agree to refrain from taking any measures that might augment the military potential of the forces under their control in the State of Jammu and Kashmir.

(For the purpose of these proposals forces under their control shall be considered to include all forces, organized and unorganized, fighting or participating in hostilities on their respective sides.)

C. The Commanders-in-Chief of the forces of India and Pakistan shall promptly confer regarding any necessary local changes in present dispositions which may facilitate the ceasefire.

D. In its discretion and as the Commission may find practicable, the Commission will appoint military observers who, under the authority of the Commission and with the cooperation of both Commands, will supervise the observance of the cease-fire order.

E. The Government of India and the Government of Pakistan agree to appeal to their respective peoples to assist in creating and maintaining an atmosphere favourable to the promotion of further negotiations.

Part II: Truce Agreement

Simultaneously with the acceptance of the proposal for the immediate cessation of hostilities as outlined in Part 1, both Governments accept the following principles as a basis for the formulation of a truce agreement, the details of which shall be worked out in discussion between their representatives and the Commission.

A

1. As the presence of troops of Pakistan in the territory of the State of Jammu and Kashmir constitutes a material change in the situation since it was represented by the Government of Pakistan before the Security Council, the Government of Pakistan agrees to withdraw its troops from that State.

2. The Government of Pakistan will use its best endeavour to secure the withdrawal from the State of Jammu and Kashmir of tribesmen and Pakistani nationals not normally resident therein who have entered the State for the purpose of fighting.

3. Pending a final solution, the territory evacuated by the Pakistani troops will be administered by the local authorities under the surveillance of the Commission.

B

1. When the Commission shall have notified the Government of India that the tribesmen and Pakistani nationals referred to in Part II, A, 2 hereof have withdrawn, thereby terminating the

situation which was represented by the Government of India to the Security Council as having occasioned the presence of Indian forces in the State of Jammu and Kashmir, and further, that the Pakistani forces are being withdrawn from the State of Jammu and Kashmir, the Government of India agrees to begin to withdraw the bulk of its forces from that State in stages to be agreed upon with the Commission.

2. Pending the acceptance of the conditions for a final settlement of the situation in the State of Jammu and Kashmir, the Indian Government will maintain within the lines existing at the moment of the cease-fire the minimum strength of its forces which in agreement with the Commission are considered necessary to assist local authorities in the observance of law and order. The Commission will have observers stationed where it deems necessary.

3. The Government of India will undertake to ensure that the Government of the State of Jammu and Kashmir will take all measures within its power to make it publicly known that peace, law and order will be safeguarded and that all human and political rights will be guaranteed.

C

1. Upon signature, the full text of the truce agreement or a communique containing the principles thereof as agreed upon public between the two Governments and the Commission, will be made public.

Part III

The Government of India and the Government of Pakistan reaffirm their wish that the future status of the State of Jammu and Kashmir shall be determined in accordance with the will of the people and to that end, upon acceptance of the truce agreement, both Governments agree to enter into consultations with the Commission to determine fair and equitable conditions whereby such free expression will be assured.

Annexure III

Shimla Agreement

Shimla Agreement on Bilateral Relations between India Pakistan signed by Prime Minister of India, Mrs. Indira Gandhi, and President of Pakistan, Mr. Z.A. Bhutto, in Shimla on July 3, 1972.

The Government of India and the Government of Pakistan are resolved that the two countries put an end to the conflict and confrontation that have hitherto marred theirs elations and work for the promotion of a friendly and harmonious relationship and the establishment of durable peace in the subcontinent, so that both countries may henceforth devote their resources and energies to the pressing task of advancing the welfare of their people.

In order to achieve this objective, the Government of India and the Government of Pakistan have agreed as follows:

(i) That the principles and purposes of the Charter of the United Nations shall govern the relations between the two countries.

(ii) That the two countries are resolved to settle their differences by peaceful means through bilateral negotiations or by any other peaceful means mutually agreed upon between them. Pending the final settlement of any of the problems between the two countries, neither side shall unilaterally alter the situation and both shall prevent the organisation, assistance or encouragement of any acts detrimental to the maintenance of peaceful and harmonious relations.

(iii) That the prerequisite for reconciliation, good neighbourliness and durable peace between them is a commitment by both the countries to peaceful co-existence, respect for each other's territorial integrity and sovereignty and non-interference in each other's internal affairs, on the basis of equality and mutual benefit.

(iv) That the basic issues and causes of conflict which have bedevilled the relations between the two countries for the last 25 years shall be resolved by peaceful means.

(v) That they shall always respect each other's national unity, territorial integrity, political independence and sovereign equality.

(vi) That in accordance with the Charter of the United Nations, they will refrain from the threat or use of force against the territorial integrity or political independence of each other.

Both Governments will take all steps within their power to prevent hostile propaganda directed against each other. Both countries will encourage the dissemination of such information as would promote the development of friendly relations between them.

In order progressively to restore and normalise relations between the two countries step by step it was agreed that:

(i) Steps shall be taken to resume communications, postal, telegraphic, sea, land including border, posts and air links including over-flights.

(ii) Appropriate steps shall be taken to promote travel facilities for the nationals of the other country.

(iii) Trade and cooperation in economic and other agreed fields will be resumed as far as possible.

(iv) Exchange in the fields of science and culture will be promoted.

In this connection delegations from the two countries will meet from time to time to work out the necessary details.

In order to initiate the process of the establishment of durable peace, both the Governments agree that:

(i) Indian and Pakistani forces shall be withdrawn to their side of the international border.

(ii) In Jammu and Kashmir, the line of control resulting from the cease-fire of December 17, 1971, shall be respected by both sides without prejudice to the recognised position of either side. Neither side shall seek to alter it unilaterally, irrespective of mutual differences and legal interpretations. Both sides further undertake to refrain from the threat or the use of force in violation of this line.

(iii) The withdrawals shall commence upon entry into force of this agreement and shall be completed within a period of 30 days thereof.

This agreement will be subject to ratification by both countries in accordance with their respective constitutional procedures, and will come into force with effect from the date on which the instruments of ratification are exchanged.

Both Governments agree that their respective heads will meet again at a mutually convenient time in the future and that in the meanwhile the representatives of the two sides will meet to discuss further the modalities and arrangements for the establishment of durable peace and normalisation of relations including the questions of repatriation of prisoners of war and civilian internees, a final settlement of Jammu and Kashmir and the resumption of diplomatic relations.

Annexure IV

Kashmir Accord (February 1975)

Agreed conclusions which led to Sheikh Mohammad Abdullah's accord with Mrs. Indira Gandhi, Prime Minister, and his subsequent assumption of office as Chief Minister in February 1975.

1. The state of Jammu and Kashmir which is a constituent unit of the Union of India, shall, in its relation with the Union, continue to be governed by Article 370 of the Constitution of India.

2. The residuary powers of legislation shall remain with the State; however, Parliament will continue to have power to make laws relating to the prevention of activities directed towards disclaiming, questioning or disrupting the sovereignty and territorial integrity of India or bringing about secession of a part of the territory of India from the Union or causing insult to the Indian National Flag, the Indian National Anthem and the Constitution.

3. Where any provision of the Constitution of India had been applied to the State of Jammu and Kashmir with adaptations and modifications, such adaptations and modifications can be altered or repealed by an order of the President under Article 370, each individual proposal in this behalf being considered on its merits; but provisions of the Constitution of India already applied to the State of Jammu and Kashmir without adaptation or modification are unalterable.

4. With a view to assuring freedom to the State of Jammu and Kashmir to have its own legislation on matters like welfare measures, cultural matters, social security, personal law and procedural laws, in a manner suited to the special conditions in the State, it is agreed that the State Government can review the laws made by Parliament or extended to the State after 1953 on any matter relatable to the

Concurrent List and may decide which of them, in its opinion, needs amendment or repeal. Thereafter, appropriate steps may be taken under Article 254 of the Constitution of India. The grant of President's assent to such legislation would be sympathetically considered. The same approach would be adopted in regard to laws to be made by Parliament in future under the Proviso to clause 2 of the Article. The State Government shall be consulted regarding the application of any such law to the State and the views of the State Government shall receive the fullest consideration.

5. As an arrangement reciprocal to what has been provided under Article 368, a suitable modification of that Article as applied to the State should be made by Presidential order to the effect that no law made by the Legislature of the State of Jammu and Kashmir, seeking to make any change in or in the effect of any provision of Constitution of the State of Jammu and Kashmir relating to any of the under mentioned matters, shall take effect unless the Bill, having been reserved for the consideration of the President, receives his assent; the matters are –

(a) the appointment, powers, functions, duties, privileges and immunities of the Governor, and

(b) the following matters relating to Elections namely, the superintendency, direction and control of Elections by the Election Commission of India, eligibility for inclusion in the electoral rolls without discrimination, adult suffrage and composition of the Legislative Council, being matters specified in sections 138, 139, 140 and 50 of the Constitution of the State of Jammu and Kashmir.

6. No agreement was possible on the question of nomenclature of the Governor and the Chief Minister and the matter is therefore remitted to the Principals.

G. Parthasarathi
Mirza Mohammad Afzal Beg
New Delhi, November 13, 1974

Annexure V

Article 370 of the Constitution of India

Temporary Provisions with respect of the State of Jammu and Kashmir.

(1) Notwithstanding anything in this Constitution:

(a) the provisions of Article 238 shall not apply in relation to the State of Jammu and Kashmir;

(b) the power of Parliament to make laws for the said State shall be limited to:

(i) those matters in the Union List and the Concurrent List which in consultation with the Government of the State, are declared by the President to correspond to matters specified in the Instrument of Accession governing the accession of the State to the Dominion of India as the matters with respect to which the Dominion Legislature may make laws for that State; and

(ii) such other matters in the said Lists as, with the concurrence of the Government of the State, the President may by order specify.

1 **Explanation:** For the purposes of this Article, the Government of the State means the person for the time being recognised by the President as the Maharaja of Jammu and Kashmir acting on the advice of the Council of Ministers for the time being in office under the Maharaja's Proclamation dated the fifth day of March, 1948.

(c) the provisions of Article (1) and of this Article shall apply in relation to this State;

(d) Such of the other provisions of this Constitution shall

apply in relation to that State subject to such exceptions and modifications as the President may by order specify:

Provided that no such order which related to the matters specified in the Instrument of Accession of the State referred to in paragraph (i) of sub-clause (b) shall be issued except in consultation with the Government of the State;

Provided further that no such order which relates to matters other than those referred to in the last preceding proviso shall be issued except with the concurrence of the Government.

2. If the concurrence of the Government of the State referred to in paragraph (ii) of sub-clause (b) of clause (1) or in the second proviso to sub-clause (d) of clause (1) or in the second proviso to sub-clause (d) of that clause be given before the Constituent Assembly for the purpose of framing the Constitution of the State is convened, it shall be placed before such Assembly for such decision as it may take thereon.

3. Notwithstanding anything in the foregoing provisions of the Article, the President may, by public notification, declare that this Article shall cease to be operative or shall be operative only with such exceptions and modifications and from such date as he may notify.

Provided that the recommendation of the Constituent Assembly of the State referred to in clause (2) shall be necessary before the President issues such a notification.

* In exercise of the powers conferred by Article 370 the President, on the recommendation of the Constituent Assembly of the State of Jammu and Kashmir, declared that as from the 17th day of November, 1952, the said Art, 370 shall be operative with the modification that for the Explanation in cl. (1) thereof, the following explanation is substituted namely.

"Explanation – For the purpose of this article, the Government of the State means the person for the time being recognized by the President on the recommendation of the Legislative Assembly of the State as the Sadar-i-Riyasat of Jammu and Kashmir, acting on the advice of the Council of Ministers of the State for the time being in office."

Annexure VI

The Armed Forces (Special Powers) Act 1958

(Act 28 of 1959)

An Act to enable certain powers to be conferred upon members of the armed forces in disturbed areas in the States' of Assam, Manipur, Meghalaya, Nagaland and Triopura and the Union Territories of Arunachal Pradesh and Mizoram,

BE it enacted by Parliament in the Ninth Years of the Republic of India as follows:

1. Short title and extent. (i) This Act may be called the Armed Forces (Special Powers) Act, 1958.

(ii) It extends to the whole of the State of Assam, Manipur, Meghalaya, Nagaland and Tripura and the Union Territories of Arunachal Pradesh and Mizoram.

2. Definitions – In this Act, unless the context otherwise requires,

(a) "armed forces" means the military forces and the air forces operating as land forces, and includes any other armed forces of the Union so operating;

(b) "disturbed area" means an area which is for the time being declared by notification under S# to be disturbed are;

(c) All other words and expressions used herein, but not defined and defined in the Air Force Act, 1950 or the Army Act 1950, shall have the meanings respectively attached to them in those Acts.

3. Power to declare areas to be disturbed areas - If, in relation to any State or Union Territory to which this Act extends the Governor of that State or the Administrator of that Union Territory, of the Central Government in either case, is of the opinion that the whole or any part of such State or Union Territory, as the case may be is in such a disturbed or dangerous condition that the use of armed forces in aid of the civil power is necessary, the Governor of that State or the Administrator of that Union Territory or the Central Government, as the case maybe, may, by notification in the Official Gazette, declare the whole or such part of such State or Union Territory to be a disturbed area.

4. Special power of the armed forces — Any commissioned officer, warrant officer, non-commissioned officer or any other person of equivalent rank in the armed forces may, in a disturbed area:

(a) if he is of opinion that it is necessary so to do for the maintenance of public order, after giving such due warning as he may consider necessary, fire upon or otherwise use force, even to the causing of death, against any person who is acting in contravention of any law or order for the time being in force in the disturbed area prohibiting the assembly of five or more persons or the carrying of weapons or of things capable of being used as weapons or of fire-arms, ammunition or explosive substances;

(b) if he is of opinion that it is necessary so to do, destroy any arms dump, prepared or fortified position or shelter from which armed attacks are made or are likely to be made or any structure used as a training camp for armed volunteers or utilised as a hideout by armed gangs or absconders wanted for any offence;

(c) arrest without warrant, any person who has committed a cognizable offence or against whom a reasonable suspicion exists that he has committed or is about to commit a cognizable offence and may use such force as may be necessary to effect the arrest;

(d) enter and search without warrant any premises to make any such arrest as aforesaid or to recover any person believed to be wrongfully restrained and confined or any property or any arms, ammunition or explosive substances believed to be unlawfully kept in such premises, and may for that purpose use such force as may be necessary.

5. Arrested persons to be made over to the police — Any person arrested and taken into custody under this Act shall be made over to the officer in charge of the nearest police station with the least possible delay, together with a report of the circumstances occasioning the arrest.

6. Protection to person acting under Act - No prosecution, suit or other legal proceeding shall be instituted, except with the previous sanction of the Central Government against any person in respect of anything done or purported to be done in exercise of the powers conferred by this Act.

7. Repeal and Saving (i) The Armed Forces (Assam and Manipur) Special Powers Ordinance, 1958, is hereby repealed.

(ii) Notwithstanding such repeal anything done or any action taken under the said ordinance shall be deemed to have been taken under this Act, as if this had commenced on the 22nd day of May, 1958.

Annexure VII

10 Year Akbar Hydari Agreement : June 1947

1. That the right of the Nagas to develop themselves according to their expressed wishes is recognised.

(a) Judicial

(i) All cases, whether civil or criminal, arising between Nagas in Naga Hills will be disposed off by duly constituted Naga Courts according to the Naga customary laws or such as may be introduced with the consent of the duly recognised Naga representative organisations save that where a sentence of transportation or death has been passed there will be right of appeal to the Governor.

(ii) In cases arising between Nagas and Non-Nagas in:

(aa) Kohima and Mokokchung Town areas, and

(ab) The judge in the neighbouring plain districts, if not Naga, will be assisted by a Naga assessor.

(b) Executive

(i) The general principle is accepted that what the Naga National Council is prepared to pay for, the Naga National Council should control. This principle will apply equally to the work done as well as the staff employment.

(ii) While the Naga District Officer will be appointed at the discretion of the Governor, sub-divisions of the Naga Hills should be administered by a sub-divisional council, who would be responsible to the District Officer for all matters falling within the latter's responsibility and to the Naga National Council for all matters falling within their responsibility.

(aa) In regard to Agriculture. The Naga National Council will exercise all the powers now vested in the District Officers.

(ab) C.W.D. The Naga National Council will take over full control.

(ac) Education and Forest Department. The Naga National Council is prepared to pay for all the services and staff.

(c) Legislative

That no law passed by the provincial or Central Legislature which would materially affect the terms of this agreement or the religious practices of the Nagas shall have legal force in the Naga Hills without the consent of the Naga National Council. In case of dispute as to whether any law did so affect this agreement, the matter would be referred by the Naga National Council to the Governor who then may direct the law in question should not have legal force in the Naga Hills pending the decision of the Central Govt.

(d) Land

That with all its resources, the Naga Hills should not be alienated to a non-Naga without the consent of the Naga National Council.

(e) Taxation

That the Naga National Council will be responsible for the imposition, collection and expenditure of land revenue and house-tax and of such other taxes as may be imposed by the Naga National Council.

(f) Boundaries

That the present administrative division should be modified so as,

(i) to bring back to the Naga Hills District all the forest areas transferred to the Sibsagar and Nowgong Districts in the past; and

(ii) to bring under one unified administrative unit, as far as possible, all Naga areas so included would be within the scope of the present proposed Agreement. No areas should be transferred out of the Naga Hills without the consent of the Naga National Council.

(g) Arms Act

The District Officer will act on the advice of the Naga National Council in accordance with the provisions of the Arms Act.

(h) Regulations

The Chin Hills Regulations and the Bengal Eastern Frontier Regulations will remain in force.

(i) Period of Agreement. The Governor of Assam as the agent of the Govt of Indian Union will have a special responsibility for a period of 10 years to ensure the due observance of this agreement; at the end of this period, the Naga National Council will be asked whether they require the above Agreement to be extended for a further period, or a new agreement regarding the future of the Naga people be arrived at.

Annexure VIII

The Shillong Agreement of November 11, 1975

The following representatives of the underground organisations met the Governor of Nagaland, Shri L.P. Singh representing the Government of India, at Shillong on 10th and 11th November, 1975.

1. Shri I. Temjenba
2. Shri S. Dahru
3. Shri Veenyiyi Rhakhu
4. Shri Z. Ramyo
5. Shri M. Assa
6. Shri Kevi Yallay

2. There was a series of four discussions. Some of the discussions were held with the Governor alone; at others, the Governor was assisted by the two Advisors for Nagaland, Shri M.

Ramunny, and Shri H. Zopianga, and Shri M.L. Kampano, Joint Secretary in the Ministry of Home Affairs. All the five members of the Liaison Committee, namely Rev. Longri Ao, Dr. M. Aram, Shri L. Lungalang, Shri Kenneth Kerhuo, and Shri Lungshim Shaiza, participated in the discussions.

3. The following were the outcome of the discussions:-

(i) The representatives of the underground organisations conveyed their decision, of their own volition, to accept, without condition, the Constitution of India.

(ii) It was agreed that the arms, now underground, would be brought out and deposited at appointed places. Details for giving effect to this agreement will be worked out between them and representatives of the Government, the security forces, and members of the Liaison Committee;

(iii) It was agreed that the representatives of the underground organisations should have reasonable time to formulate other issues for discussion for final settlement.

Dated Shillong,

November 11, 1975.

Sd/-(I. Temjenba)

Sd/-(S. Dahru)

Sd/-(Z. Ramyo)

Sd/-(M.Assa)

Sd/-(Kevi Yallay)

On behalf of the representative of underground organisations.
Representative of the Government of India.

Sd/-(LP Singh)

On behalf of the Government of India.

Supplementary Agreement of January 5, 1976

Implementation of Clause II of the Shillong Accord of November 11, 1975

1. It was decided that the collection of arms, initially at collection centres, would commence as early as possible, and will be completed by 25th January, 1976. Initial places of collection to be decided through discussion between Commissioner, representatives of underground organisations and the members of the Liaison Committee.

2. Once all arms are collected, these will be handed over to Peace Council team at the respective places of collection.

3. Peace Council team will arrange to transport the arms from collection centres to Chedema peace camp and arrange guards, etc., for safe custody of arms.

4. Similar arrangement at agreed place/places will be made in Manipur with the concurrence of the Manipur Government.

5. The underground may stay at peace camps to be established at suitable places, and their maintenance will be arranged only by the Peace Council. Any voluntary contribution from any source will be made to the Peace Council who will utilise the fund according to necessity.

Sd/-

(L.P. Singh)

Governor

1. Sd/-

(Biseto Medom Keyho)

2. Sd/-

(Pukrove Nakru)

3. Sd/-

(Z. Ramyo)

4. Sd/-

(I. Temjenba)

Shillong

Dated January 5, 1976

Index

A

Afghanistan, 11, 63, 91, 92
AIDS, 79
Akbar, 158
Alexander the Great, 121
Algeria, 17
Andhra Pradesh, 101
Arrival, 57-64
  area familiarization, 61-63
  extreme caution, 63-64
  logistic support, 59-61
  protection arrangements, 58-59
  reaction of people, 57-58
Assam, 93, 123, 197
Axis of Maintenance, 190-196

B

Baldwin, Hansan W, 11
Bhutto, Benazir, 20
Biao, Lin, 160
Birbal, 158
Bonnet, Colonel Gabriel, 12
Border Security Force (BSF), 43, 82, 89
Brar, Lt Gen KS, 37

C

Cambodia, 16
China, 16, 17
Clevald, Martin Van, 15, 209
Central Reserve Police Force (CRPF), 82, 89, 113
Chamberlain, Sir N, 177
Chanakya, 177
Communications, 185-189
Cyprus, 17

D

Dar, Abdul Majid, 98
Doda, 41, 51, 53, 58, 67, 79, 179, 190, 191

E

Eelam People's Revolutionary Liberation Front (EPRLF), 125
Egypt, 192
Establishing a Post, 65-74
  alarm, 71-72
  arrangements for guards, 70-71
  axis of maintenance, 73
  barriers, 70
  consensus, 73-74
  defence capability : analysis and layout, 68-69
  interrogation room, 72
  nets, 70
  security lights, 69-70
  selecting a site, 65-66
  taking over of accommodation, 66-68
  vehicle park, 72
  visitors' room, 72-73

F

Fall, Bernard B, 12
Fauji, 205
Farooq, Mirwaiz Maulvi, 113
First Impact, 75-81
  action, 80
  catch and release, 80-81
  first impression, 76-77
  meet all residents, 78-79
  participation in festivals, 81
  points that can be conveyed to the people, 79-80
  speak to elders, 77-78
Flexibility, 150-155
Frank Kitson, 10
Fuller, Maj Gen John Frederick Charles, 108

G

Gallagher, James J, 11
Gardner, John W, 200
Goff, Stanley, 116, 118
Ground, 197-199

H

Harkat-Ul –Ansar, 44
Hindus, 58, 94

Hizbul Mujahideen, 28, 41
Human Rights, 100-105
Hungary, 39

I

India, 16, 17, 19, 20, 85, 98, 101
  Constitution of, 101
Indian Peacekeeping Force (IPKF), 57, 126, 138, 190
Indonesia, 17
Initiative, 158-165
  OP ADVENTURE PATROL KATAR, 162-165
Intelligence, 173-184
  OP Rheni, 179
Interaction with Civil Administration, 85-91
  joint action committee, 88-91
  meet administrative staff, 86-87
  unified command, 87-88
Interaction with Paramilitary Forces, 82-84
  gauge effectiveness, 83-84
  impression of faith, 84
  meet commanders at all levels, 82-83
  mutual respect, 83
Inter-Services Intelligence (ISI), 20, 142
Iran-Iraq War, 11

J

Jagmohan, 204
Jammu & Kashmir, 20, 21, 41, 85, 101, 102, 155, 161, 179, 190, 204, 210

K

Kashmir, 11, 20, 79, 94, 200
Keagan, John, 24
Kenya, 17
Keppel, David, 33
Khajooria, M M, 101

L

Laos, 15, 16
Lebanon, 16
Liberation Tigers of Tamil Eelam (LTTE), 19, 47, 51, 94, 95, 125, 126, 128
Linhuang, Wei, 160
Leadership, 24-36
  all kinds of reasons, 31-32
  believe in fight, 30-31
  confidence, 34-35
  courage, 27-28
  humour, 32-33
  initiative, 25
  lead by example, 26-27
  morality, 29-30
  patience, 28-29
  physical and mental resilience, 35-36
  professional competence, 33-34
  vision, 36
LIC (Low Intensity Conflict), 10-22, 24, 25, 26, 27, 29, 30, 32, 33, 34, 35, 36, 43, 47, 50, 51, 53, 54, 55, 56, 57, 58, 71, 74, 76, 82, 84, 85, 92, 100, 101, 104, 108-111, 112, 113, 121, 122, 123, 124, 131, 132, 141, 142, 150, 151, 159, 166, 167, 173, 175, 176, 179, 185, 186, 187, 190, 197, 200, 203, 204, 205, 206, 209, 210
  defining of, 10-12
  environment of Indian, 19-22
    Pakistan factor, 19-20
    Role of Company Commander, 20-22
  facets of, 15-18
  principles of, 108-111, 123

M

Madras, 57
Maheshwari, Anil, 95
Manipur, 20, 21, 22, 94, 147, 169, 197
Media Management, 204-206
Minh, Ho Chi, 200, 201
Mintang, Kuo, 161
Mobility and Manoeuvre, 140-149
  Genesis of Ambush Plan 'Dhanush', 142-146
  OPERATION KACHAI, 146-149
Mother Teresa, 81
Moral Ascendancy, 200-203
  twelve recommendations, 201-203
Mujahideen, 18
Mujhawar, Havildar B.H, 43, 44
Muslims, 57, 58

N

Nagaland, 20, 21, 22, 94, 197
  National Socialist Council of Nagaland (NSCN), 94, 146, 147, 149, 169
Naik Ajay Kumar, 53
Namibia, 18

O

Offensive Action, 121-130
  Operation Kontakkarankulam, 125-126
  Operational Plan, 126-130
OP ADVENTURE PATROL KATAR, 162-165
OP BAJRANG, 20, 48
OP HIFAZAT, 169-172
OP JACKAL, 156-157
OP KACHAI, 146-149
OP MERRY LANE, 135-139
OP PAWAN, 20, 25, 26, 51, 57, 134, 187, 190
OP RHINO, 20

P

Pakistan, 19, 20, 43, 78, 79, 97, 102, 142, 204
Pakistan Occupied Kashmir (POK), 145
Palestine, 17
Param Vir Chakra, 205
People's Liberation Army (PLA), 160, 161
Philippines, 159
Preparations for Move, 37-56
  deficiency management, 54
  induction training, 42-47
  intelligence collection, 47-49
  modification of equipment, 50-54
  physical conditioning, 49-50
  psychological, 39-41
  representation of supporting arms, 55
  screening of counter-insurgency films, 56
Psychological Operation, 92-99
  achieving goodwill, 98-99
  developmental projects, 97-98
  futility of insurgency, 96
  insurgent weaknesses vs security forces' strengths, 95-96
  level of support, 93-94
  psychology of the insurgent, 94-95
  publicize good work, 96-97
Punjab, 93, 101, 200

R

Rao, Gen KV Krishna, 101

S

Sardeshpande, Lt Gen SC, 25, 94, 190
Second World War, 29, 41, 123, 152, 191
Security, 166-172
  OP HIFAZAT, 169-172
Selection and Maintenance of Aim, 112-120
  pitched battles with 'Charlies', 116-120
South Africa, 15, 18
Soviet Union (USSR), 18, 91, 92, 152
Srinagar, 51, 57, 82
Sri Lanka, 19, 47, 69, 125
Surprise, 131-139
  OP Merry Lane, 135-139

T

Tamils, 57
Thapa, Havildar Bhim Bahadur, 125, 128, 129
Tomorrow, 209-212
Tzu, Sun, 197, 198

U

United Liberation Front of Assam (ULFA), 48, 93, 123, 150, 151
United States (US), 11, 20, 75, 88, 92, 109, 159
  forces of, 75, 91, 109, 159

V

Vietnam, 11, 30, 75, 91, 92, 116, 168, 169, 200

W

Wirsing, Robert, 20

Y

Yom Kippur War, 11

Z

Zhou, En Lai, 161
Zedong, Mao, 158, 161